EDUCATION IN A COMPETITIVE AND GLOBALIZING WORLD

COMPUTER-ENABLED MATHEMATICS: INTEGRATING EXPERIMENT AND THEORY IN TEACHER EDUCATION

EDUCATION IN A COMPETITIVE AND GLOBALIZING WORLD

Additional books in this series can be found on Nova's website under the Series tab.

Additional E-books in this series can be found on Nova's website under the E-book tab.

MATHEMATICS RESEARCH DEVELOPMENTS

Additional books in this series can be found on Nova's website under the Series tab.

Additional E-books in this series can be found on Nova's website under the E-book tab.

EDUCATION IN A COMPETITIVE AND GLOBALIZING WORLD

COMPUTER-ENABLED MATHEMATICS: INTEGRATING EXPERIMENT AND THEORY IN TEACHER EDUCATION

SERGEI ABRAMOVICH

Nova Science Publishers, Inc.
New York

For permission to use material from this book please contact us:
Telephone 631-231-7269; Fax 631-231-8175
Web Site: http://www.novapublishers.com

Additional color graphics may be available in the e-book version of this book.

LIBRARY OF CONGRESS CATALOGING-IN-PUBLICATION DATA

Abramovich, Sergei.
Computer-enabled mathematics : integrating experiment and theory in teacher education / author, Sergei Abramovich.
p. cm.
Includes index.
ISBN 978-1-61122-784-0 (hardcover)
1. Mathematics teachers--Training of. 2. Mathematics--Study and teaching (Secondary)--Data processing. 3. Mathematics--Computer-assisted instruction. 4. Electronic data processing--Study and teaching (Secondary)
I. Title.
QA10.5.A225 2010
510.71--dc22
2010041357

Published by Nova Science Publishers, Inc. † New York

CONTENTS

PREFACE

This book is based on the author's experience in teaching a computer-enhanced capstone course for prospective teachers of high school mathematics (referred hereafter to as teachers). The book addresses core recommendations by the Conference Board of the Mathematical Sciences (2001)—an umbrella organization consisting of seventeen professional societies in the United States—regarding the mathematical preparation of teachers. According to the Board, the concept of a capstone course in a mathematics education program includes teachers' learning to use commonly available and user-friendly software tools with the goal to reach a certain depth of the mathematics curriculum through appropriately designed computational experiments. In turn, the notion of experiment in the teaching of mathematics sets up a path toward enhancing the "E" component of teachers' literacy in the STEM (science, technology, engineering, mathematics) disciplines because the integration of experimental and theoretical approaches to mathematical learning has the potential to shape engineering mindset of the teachers (Katehi, Pearson, and Feder, 2009).

An experimental approach to mathematics draws on the power of computers to perform numerical computations and graphical constructions, thereby enabling easy access to mathematical ideas and objects under study. The approach includes one's engagement in recognizing numerical patterns formed by modeling data and formulating properties of the studied objects through interpreting behavior of their geometric representations. This makes it possible to balance formal and informal approaches to mathematics allowing teachers to learn how the two approaches complement each other.

Several computer applications are used throughout the book. One application is a spreadsheet. Nowadays, facility at creating a spreadsheet is required in many entry-level positions for high school graduates and, to some extent, the intelligent use of software is expected from educators across the spectrum of disciplines. Another application is *The Geometer's Sketchpad* (*GSP*)—dynamic geometry software commonly available in North American schools and elsewhere in the world. Like spreadsheets, the *GSP* includes many features conducive to experimentation with mathematical concepts as well as generating insight and understanding. The book also incorporates *Maple*—a computer algebra system for mathematical modeling—that makes it possible to use symbolic computation as a way of reducing repetitive, lengthy, and error-likely paper-and-pencil work and, instead, emphasizing conceptual growth and the development of formal reasoning skills. Finally, the book has a strong focus on the use of the *Graphing Calculator 3.5* (*GC*)—software produced by Pacific Tech—that supports and facilitates the use of geometric method by enabling the construction of graphs from any two-variable equation or inequality. In particular, such use of the *GC* encourages digital fabrication—the process of using a computer to create a digital design with the goal to translate it into a physical object (Gershenfeld, 2005).

The book's content is a combination of mathematical concepts typically associated with secondary problem-solving curriculum and their extensions into the tertiary curriculum made possible by the use of technology. By the same token, using technology allows one to see how the roots of higher concepts penetrate mathematical ideas at the elementary level. Towards this end, the first chapter uses one of the most basic objects of elementary school mathematics—the multiplication table—as a window to the concepts of algebra, discrete mathematics, and calculus. Typically, in arithmetic, the multiplication table is introduced as a static medium the only "mission" of which is to record and store in a strict order various multiplication facts for the purpose of memorization. However, the table may be used also as a mathematical object (like matrix) to which different operations can be applied and geometric interpretations of those operations can be developed. Such a dynamic perspective on the multiplication table, when enhanced by the use of interactive spreadsheets, enables a variety of mathematical investigations that span from the summation of arithmetic sequences to deciding the convergence of series.

The second chapter is devoted to the study of algebraic equations with parameters through the so-called locus approach that goes back to Descartes whose wondrous insight led to realization that one-to-one correspondence between an algebraic equation in two variables and a curve in the coordinate plane can be established. Through interactive experimentation with graphs in the context of the *GC*, the locus approach makes it possible to conceptualize and analytically formulate many properties of equations that the graphs represent. Once these properties have been established, they can be formally demonstrated using the language of algebra and then verified through a new experiment.

In the third chapter, spreadsheet-based computing applications are utilized to motivate the use of algebraic inequalities and associated proof techniques. These applications deal with the construction of computationally efficient environments, which, in turn, can be utilized as generators of new problems solvable, again, through the use of inequalities. Such an approach of utilizing mathematical concepts as emerging tools in computing applications shifts the focus of activities from using computers for solving inequalities to using inequalities for improving computational efficiency of computers.

The fourth chapter introduces the notion of geometric probability and integrates mathematical machinery that differs in complexity from proper fractions to definite integrals. Supported by the appropriate context, the geometric perspective on calculating probabilities makes it possible to uncover hidden properties of fractions and use functions and their graphs to give meaning to typically overlooked arithmetical phenomena. The chapter also demonstrates how computational experiments in the context of spreadsheets can be used to confirm theoretical calculations made possible by integral calculus enhanced by the *GC*.

The fifth chapter introduces various concepts of enumerative combinatorics, both elementary and advanced, by using the unity of context, numeric modeling, visualization, symbolic computation, and formal mathematics. Stemming from semantically uncomplicated problems, these concepts are formulated in terms of difference equations—mathematical models of discrete dynamical systems found in many engineering applications. In turn, a spreadsheet is used to numerically model these equations, thereby creating a numeric environment for recognizing patterns that numbers with combinatorial meaning generate.

Furthermore, the numerical modeling approach enables the discovery of non-trivial connections between different combinatorial concepts, something that would not have been possible in the absence of computers.

The sixth chapter sheds new light on how one can integrate context, mathematics, historical perspectives, and computers in a capstone course. Here a spreadsheet, the *GC*, and the *GSP* come together to demonstrate the potential of technology to deepen one's insight into a number of historically significant explorations that span from antiquity to the 19th century. Furthermore, a focus on historical perspectives makes it possible to revisit ideas and concepts introduced earlier in the book.

The seventh chapter deals with trigonometry—a part of school mathematics where, traditionally, the use of a computer has been limited to the calculation of the values of circular and arc functions and the construction of their graphs. Drawing on the notion of equivalence, the chapter incorporates the *GC*-based computational experiments to explain several trigonometry-specific phenomena and to develop understanding of hidden properties of solutions of trigonometric equations depending on parameters. Here, once again, the greatness of geometrization as a method is emphasized by demonstrating how certain ideas of trigonometry can be understood from a geometric perspective.

The eighth chapter revisits ideas about computational problem solving and modeling introduced elsewhere (Abramovich, 2010) and elevates these ideas at a mathematically and computationally higher level by drawing on a variety of tools discussed throughout the book including arithmetic sequences, polygonal numbers, algebraic inequalities, quadratic functions, and systems of equations. Whereas the main software tool used in this chapter is a spreadsheet, the *GC* and *Maple* are used also as appropriate. The focus is on the applied nature of mathematical concepts and on a possibility of using a spreadsheet first as an agent, then as a consumer, and, finally, as an amplifier of problem-solving activities.

The last chapter plays the role of appendix by providing programming details for most of the spreadsheets used in the book. From it, teachers are expected to gain a technical expertise in designing spreadsheet-based learning environments. As to other software tools that the book utilizes, the corresponding technical details are discussed concurrently with the tools' use in support of computational experiments and geometrization of algebraic ideas.

To conclude, note that the book attempts to contribute to the preparation of qualified teachers as "the best way to raise [average] student achievement" (Conference Board of the Mathematical Sciences, 2001, p. 3). Such qualification is also a crucial factor in realizing full potential of capable students by appreciating and nurturing their creative ideas. The material included in the book is mostly original in terms of its emphasis on the experimental approach to school mathematics and it is based on a number of journal articles published by the author in the United States and elsewhere. Mathematics educators interested in integrating commonly available software tools in a capstone course that follows current recommendations for teacher preparation will find this book useful. The book can also be of interest to practicing teachers (and their students alike) who want to enhance their knowledge of secondary mathematics and computer applications in the context of experimentation with mathematical ideas.

Chapter 1

THE MULTIPLICATION TABLE FROM AN ADVANCED STANDPOINT

The fact that in actual practice counting is limited is not relevant; an abstraction is made from it. It is with this indefinitely prolonged sequence that general theorems about numbers have to deal.

— Alexandrov (1963, p. 16)

1. INTRODUCTION

Current standards for teaching school mathematics (National Council of Teachers of Mathematics, 2000) and recommendations for teacher preparation in mathematics (Conference Board of the Mathematical Sciences, 2001) emphasize the importance of making connections among different mathematical ideas, concepts, and curricular topics as a means of providing rich instructional and learning opportunities. It has been argued that "core mathematics major courses can be redesigned to help teachers make insightful connections between the advanced mathematics they are learning and the high school mathematics they are teaching" (ibid, p. 39). Using the multiplication table as a background, this chapter provides several teaching ideas regarding connections that can be made among concepts typically encountered in the study of algebra, geometry, discrete mathematics, functions, and calculus. A common thread that permeates those ideas is the use of "the increasingly sophisticated technological tools that permit more computationally involved applications and can give insights into theory" (ibid, p. 37). By focusing on the numerical approach to mathematics in a capstone course made possible by the use of technology, one can develop the teachers' appreciation of how concrete and abstract representations of mathematical concepts can be bridged effectively across the curriculum.

This chapter consists of problems with solutions and propositions with proofs. Such a distinction makes it possible to show how one's engagement in problem solving builds a foundation for the development of mathematical propositions. In the words of Pólya[1] (1973), "Good ideas are based on past experience and formerly acquired knowledge" (p. 9). A few

[1] George Pólya (1887-1985)—a Hungarian born American mathematician known for his outstanding contributions to the fields of classical analysis and mathematics education.

basic propositions are presented first in the form of algebraic formulas used as tools in solving problems of a geometric nature. This highlights another important aspect of the teaching ideas presented in this chapter (and elsewhere in the book)—the didactic significance of geometric roots of algebraic propositions. Put another way, the chapter emphasizes the important role of intuition and context in motivating the growth of mathematical concepts. Nowadays, computer-enhanced and experimentally based approaches to mathematics facilitate and support that kind of teaching and learning of the subject matter.

2. Basic Summation Formulas

Although the summation formulas that appear in this section can be shown to be motivated by solving concrete problems, for the sake of brevity and simplification of the future exposition of ideas they will be introduced in a formal way.

Proposition 1. The sum $S_n = a_1 + a_2 + ... + a_n$ of the first n terms of the arithmetic series $\{a_i\}$—a numeric sequence with a constant difference d between two consecutive terms—can be found as

$$S_n = \frac{a_1 + a_n}{2} n \qquad (1)$$

Proof. By adding the n terms twice yields

$$2S_n = (a_1 + a_n) + (a_2 + a_{n-1}) + ...(a_i + a_{n-i+1}) + ... + (a_n + a_1)$$

Noting that for $1 < i < n$

$$a_i + a_{n-i+1} = a_1 + d(i-1) + a_1 + d(n-i) = a_1 + a_1 + d(n-1) = a_1 + a_n$$

one has $2S_n = (a_1 + a_n) \cdot n$ whence formula (1).

Corollary 1. The sum of the first n counting numbers can be found through the formula

$$1 + 2 + 3 + ... + n = \frac{n(n+1)}{2} \qquad (2)$$

Proof. The sequence of consecutive counting numbers is an arithmetic sequence with the difference $d = 1$. Thus, substituting 1 for a_1 and n for a_n in formula (1) results in formula (2).

Corollary 2. The sum of the first n odd numbers can be found through the formula

$$1 + 3 + 5 + ... + 2n - 1 = n^2 \qquad (3)$$

Proof. The sequence of consecutive odd numbers is an arithmetic sequence with the difference $d = 2$. Thus, substituting 1 for a_1 and $2n - 1$ for a_n in formula (1) results in formula (3).

Remark 1. The sum of the first n counting numbers is called the n-th triangular number or the triangular number of rank n. These numbers will be used in multiple contexts throughout the book and denoted as t_n. Formula (2) is a closed formula for triangular numbers. Substituting $n - 1$ for n in formula (2) yields $t_{n-1} = \frac{(n-1)n}{2}$. Furthermore,

$$t_{n-1} + n = \frac{(n-1)n}{2} + n = \frac{n(n+1)}{2} = t_n .$$

Therefore, the relation $t_n = t_{n-1} + n$ is a recursive formula for triangular numbers.

Proposition 2. The sum of the first n squares of counting numbers can be found through the formula

$$1^2 + 2^2 + 3^2 + \ldots + n^2 = \frac{n(n+1)(2n+1)}{6} \quad (4)$$

Proof. It follows from formula (3) that the difference between two consecutive square numbers is not a constant and therefore formula (1) cannot be used in this case. Instead, the method of mathematical induction, referred to by Pólya (1954) as "the demonstrative phase" (p. 110), or, alternatively, "transition from n to $n + 1$" (p. 112), will be used to demonstrate that formula (4) is true for all n. The first step of the method is to show that formula (4) is true for $n = 1$. Indeed, when $n = 1$, both sides of (4) are equal to one. The second step is to assume that formula (4) holds true (in other words, to make the so-called inductive assumption) and then show that after replacing n by $n + 1$ it remains true (verifying transition from n to $n + 1$), that is

$$1^2 + 2^2 + 3^2 + \ldots + n^2 + (n+1)^2 = \frac{(n+1)(n+2)(2n+3)}{6}.$$

Indeed,

$$\begin{aligned}
&1^2 + 2^2 + 3^2 + \ldots + n^2 + (n+1)^2 \\
&= \frac{n(n+1)(2n+1)}{6} + (n+1)^2 \\
&= \frac{(n+1)(n(2n+1) + 6(n+1))}{6} \\
&= \frac{(n+1)(2n^2 + 7n + 6)}{6} = \frac{(n+1)(n+2)(2n+3)}{6}.
\end{aligned}$$

This completes the proof.

Proposition 3. The sum of the first n cubes of counting numbers can be found through the formula

$$1^3 + 2^3 + 3^3 + ... + n^3 = (\frac{n(n+1)}{2})^2 \tag{5}$$

Proof. Formula (5) can also be proved by mathematical induction (its geometric interpretation is discussed in section 5 of this chapter). When $n = 1$, both sides of (5) are equal to one. Assuming that (5) is true, transition from n to $n + 1$ can be carried out as follows

$$1^3 + 2^3 + 3^3 + ... + n^3 + (n+1)^3 = (\frac{n(n+1)}{2})^2 + (n+1)^3$$
$$= \frac{(n+1)^2}{4}(n^2 + 4n + 1)$$
$$= \frac{(n+1)^2(n+2)^2}{4} = (\frac{(n+1)((n+1)+1)}{2})^2.$$

This completes the proof.

3. On the Geometric Meaning and Inductive Conjecturing of Formula (4)

Because the difference $(n+1)^2 - n^2 = 2n+1$ is a variable quantity that depends on n, consecutive squares do not form an arithmetic series the terms of which can be paired to make equal sums. In order to develop a method of summation of the first n squares of counting numbers, that is, formula (4), consider a special case of $n = 4$

$$1^2 + 2^2 + 3^2 + 4^2 \tag{6}$$

According to formula (3),

$$1^2 = 1,\ 2^2 = 1+3,\ 3^2 = 1+3+5,\ 4^2 = 1+3+5+7.$$

Therefore, by analogy with the method used in proving formula (1), that is, extending what sometimes is referred to as Gauss's idea[2] of summation (Pólya, 1981) to the case of consecutive squares, one can represent sum (6) in the following three ways

$$1 + (1 + 3) + (1 + 3 + 5) + (1 + 3 + 5 + 7)$$

[2] Carl Friedrich Gauss (1777-1855, Germany)—one of the greatest mathematicians in the history of mankind.

$$1 + (7 + 3) + (5 + 1 + 3) + (5 + 1 + 3 + 1)$$

$$7 + (1 + 3) + (3 + 5 + 1) + (3 + 5 + 1 + 1)$$

so that:

(i) each of the three sums of ten numbers arranged into four groups includes four ones, three threes, two fives, and one seven;
(ii) each of the ten vertical sums of the corresponding three numbers from each of the three sums has the same value, 9, across all such sums.

That is, sum (6), the alternative representation of which is

$$(1 + 1 + 1 + 1) + (3 + 3 + 3) + (5 + 5) + 7, \qquad (7)$$

when repeated three times is equal to the product $10 \cdot 9$. Consequently, sum (6) is equal to one-third of this product, that is, $1^2 + 2^2 + 3^2 + 4^2 = \dfrac{10 \cdot 9}{3} = 30$. Whereas the meaning of the first factor is rather clear (if one uses (7) to count the number of summands), that is, $10 = 1 + 2 + 3 + 4$, the meaning of the second factor, 9, is less obvious and it can be revealed through the following modification of the above three sums

$$1 + (1 + 3) + (1 + 3 + 5) + (1 + 3 + 5 + 7)$$

$$4 + (4 + 3) + (4 + 3 + 2) + (4 + 3 + 2 + 1)$$

$$4 + (4 + 3) + (4 + 3 + 2) + (4 + 3 + 2 + 1)$$

or

$$(1 + 1 + 1 + 1) + (3 + 3 + 3) + (5 + 5) + 7$$

$$(4 + 4 + 4 + 4) + (3 + 3 + 3) + (2 + 2) + 1$$

$$(4 + 4 + 4 + 4) + (3 + 3 + 3) + (2 + 2) + 1.$$

One can see that $9 = 4 + 4 + 1$, that is, $9 = 2 \cdot 4 + 1$. Furthermore, one can see that each of the last two sums has the form of sum (6):

$$4 \cdot 4 + 3 \cdot 3 + 2 \cdot 2 + 1 \cdot 1 = 1^2 + 2^2 + 3^2 + 4^2$$

In order to give a geometric interpretation to the above arithmetical experimentation with numbers, pictorial representations of sums (6) and (7) can be introduced in the form of the towers shown in Figures 1.1 and 1.2, respectively.

Combining two towers representing sum (6) and one tower representing sum (7) results in the rectangle pictured in Figure 1.3. Just as it was shown in the domain of arithmetic, geometrically, the number of blocks comprising this rectangle is three times as much as the number of blocks comprising the tower shown in Figure 1.1 (or Figure 1.2), can be found as the product $(1+2+3+4)(2\cdot 4+1)$. Furthermore,

$$1+2+3+4=\frac{4\cdot(4+1)}{2}.$$

Therefore,

$$1^2+2^2+3^2+4^2=\frac{1}{3}\cdot[\frac{4\cdot(4+1)}{2}\cdot(2\cdot 4+1)]=\frac{4\cdot(4+1)\cdot(2\cdot 4+1)}{6}.$$

In much the same way, one can develop the relations

$$1^2+2^2+3^2=\frac{3\cdot(3+1)\cdot(2\cdot 3+1)}{6}$$

and

$$1^2+2^2=\frac{2\cdot(2+1)\cdot(2\cdot 2+1)}{6}.$$

Generalizing from the above three special cases to the sum of the first n squares of counting numbers results in formula (4).

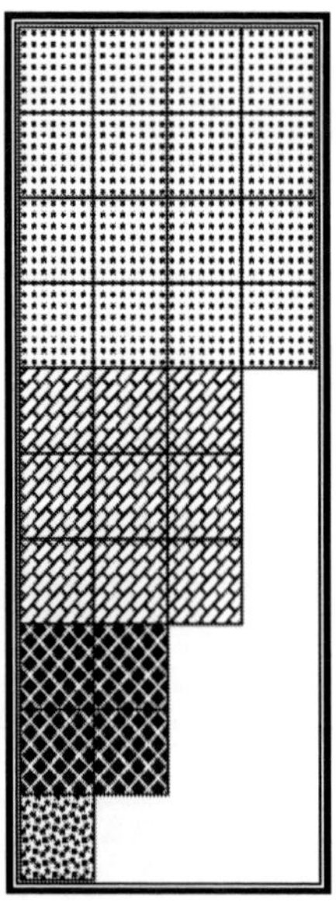

Figure 1.1. A pictorial representation of sum (6).

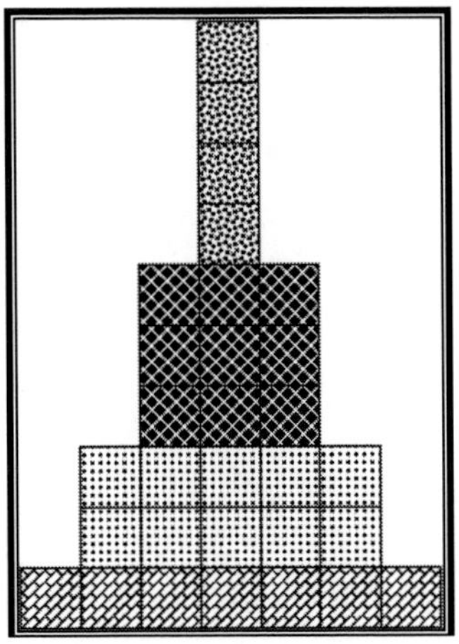

Figure 1.2. A pictorial representation of sum (7).

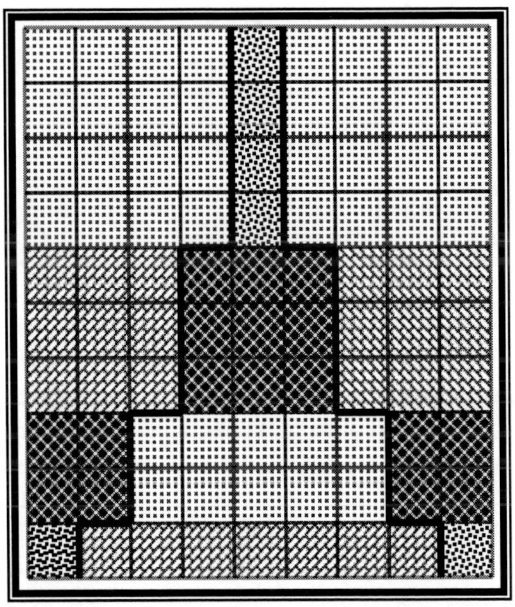

Figure 1.3. The sum of the first four squares of counting numbers increased three-fold.

4. On the Deficiency of Inductive Reasoning

Mathematical induction proof differs from inductive reasoning in the following significant way: The former argument provides rigor and the latter argument may lead to an incorrect generalization. In other words, "induction is never conclusive" (Pólya, 1954, p. 171). For example, consider the problem of finding the number of segments of *different* non-integer lengths within a $n \times n$ grid (cf. National Council of Teachers of Mathematics, 2000, p. 266).

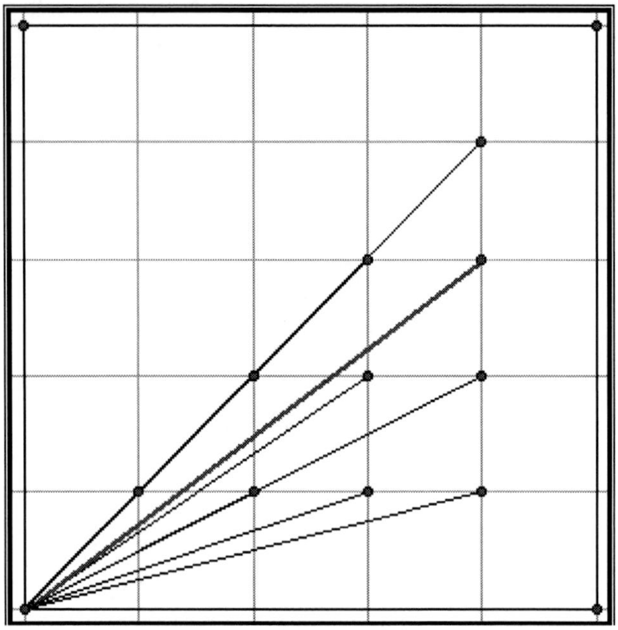

Figure 1.4. Diagonally connecting dots on a geoboard.

As shown in Figure 1.4, when n = 1 we have one such segment, when n = 2 we have three such segments, when n = 3 we have six such segments and it appears that when n = 4 we have 10 such segments. In other words, the number of segments appears to be represented each time by a triangular number so that one may conjecture (generalize) there are t_n segments of different non-integer lengths within a $n \times n$ grid. However, in the case n = 4 one of the segments has length five ($3^2 + 4^2 = 5^2$). Thus, the emergence of the elements of Pythagorean triples among the segments defies this conjecture obtained by inductive reasoning. A sudden collapse of a seemingly plausible generalization also shows that whereas multiple examples that support a conjecture may not be considered a proof, a single counterexample is sufficient to defy the conjecture.

Another example that demonstrates the deficiency of reasoning by induction can be derived from the (geometric) analysis of the method used in developing formula (4). Observing Figure 1.3, one may note that area of the rectangle built out of the three towers is equal to $9 \times 10 = 90$ square units, an observation already made through a numerical experimentation. At the same time, perimeter of the rectangle is equal to $2 \times (9+10) = 38$ linear units. Other possible arrangements of 90 square units (individual blocks shown in Figure 1.3) would result in rectangles with perimeters greater than 38 linear units. For example, $90 = 6 \times 15$ and $2 \times (6+15) = 42 > 38$. If one constructs similar rectangles for other sums of consecutive squares, the following numbers for area and perimeter would be found: in the case $1^2 + 2^2$ we have area 15 and perimeter 16—the smallest possible for that area; in the case $1^2 + 2^2 + 3^2$ we have area 42 and perimeter 28—the smallest possible for that area; in the case $1^2 + 2^2 + 3^2 + 4^2 + 5^2$ we have area 165 and perimeter 52—the smallest possible for that area. These observations may lead to the following inquiry into the method of developing formula (3): Does this method have a hidden connection to the property of rectangles to possess the smallest perimeter, given the area? That is, the use of inductive reasoning might result in the conclusion that in constructing rectangles out of three towers one always arrives at the rectangle with the smallest perimeter, regardless of the value of n in formula (4).

To clarify, note that, in general, the number of unit squares included into three towers, alternatively, the area $A(n)$ of the resulting rectangle, is equal to $A(n) = \frac{n(n+1)(2n+1)}{2}$. Let x be a side length of one of such rectangles; then the other side length and perimeter of the rectangle are equal, respectively, to $\frac{n(n+1)(2n+1)}{2x}$ and $P(x,n) = 2x + \frac{n(n+1)(2n+1)}{x}$. Using the Arithmetic Mean—Geometric Mean inequality (described in detail in Chapter 3), one can estimate $P(x,n)$ from below as follows:

$$P(x,n) = 2x + \frac{n(n+1)(2n+1)}{x} \geq 2\sqrt{2n(n+1)(2n+1)}.$$

On the other hand, the perimeter of the rectangle (shown in Figure 1.3 for $n = 4$) is equal to $f(n) = 2[\frac{n(n+1)}{2} + 2n + 1] = n^2 + 5n + 2$. Surprisingly, the values of the functions $f(n)$ and $g(n) = CEILING[2\sqrt{2n(n+1)(2n+1)}]$, as can be shown by using a spreadsheet, coincide for $n = 2$, 3, 4, and 5, thereby, confirming our earlier observations. Here, $CEILING(x)$ is the smallest integer greater or equal to x. Note that for rectangles with whole number sides the inequality $P(x,n) \geq CEILING[2\sqrt{2n(n+1)(2n+1)}]$ holds true. However, whereas $f(n)$ is a quadratic function, the function $g(n)$ grows as $n^{3/2}$. For example, whereas $f(10) = 152$, we have $g(10) = 136 = 2 \cdot 68 = 2 \cdot (33 + 35)$ and $A(10) = 1155 = 33 \cdot 35$. These calculations may serve as a counterexample that defies the conjecture regarding perimeters of rectangles used to find the sum of the first n squares—as it turned out, in general, those rectangles do not have the smallest perimeter.

5. Checkerboard Problem and its Different Extensions

Checkerboard Problem. How many different rectangles can be found on the $n \times n$ checkerboard?

In his famous book on mathematical problem solving, Pólya (1973) expressed the following thoughts about teaching: “The first rule of teaching is to know what you are supposed to teach. The second rule of teaching is to know a little more than what you are supposed to teach” (p. 173). With the second rule in mind, different extensions of the checkerboard problem, supported by a spreadsheet as an interactive computational medium conducive to a variety of experiments with numbers, will be discussed below. One such extension includes finding on the checkerboard the number of squares and rectangles (in particular, squares) with special properties (e.g., those having at least one side measured by an odd number). Another possible extension of the checkerboard problem is to find the number of prisms and the number of cubes within the $n \times n \times n$ cube. These extensions, motivated by the above Pólya's maxim and driven by numerical computations made possible by the use of interactive multiplication tables, would allow for the comparison of the rates of growth of the cardinal numbers of each set of the geometric figures. Towards this end, connections between secondary and tertiary mathematical concepts—one of the main objectives of a capstone course—will be established. The programming details of the spreadsheet-based multiplication table of a variable size that incorporates conditional formatting features are discussed in Chapter 9.

To begin note that the checkerboard problem supports the problem solving standard of the Principles and Standards for School Mathematics (National Council of Teachers of Mathematics, 2000, p. 335), where it is shown that the total number of rectangles within the $n \times n$ checkerboard (for which we will use the notation $N_{rects}[n]$) can be found by adding up all numbers in the corresponding $n \times n$ multiplication table (n = 8 in Figure 1.5). In turn, the sum of all numbers in such a table is equal to the square of the sum of the first n counting numbers. Indeed, each number in row k, $1 \leq k \leq n$, of the multiplication table is k times as much as the corresponding number in row one of the table. As the sum of numbers in the first row is equal to the sum of the first n counting numbers, we have

$$\begin{aligned} &1 \cdot (1+2+...+n) + 2 \cdot (1+2+...+n) + ... + n \cdot (1+2+...+n) \\ &= (1+2+...+n)(1+2+...+n) \\ &= (1+2+...+n)^2. \end{aligned}$$

The use of formula (2) yields

$$N_{rects}[n] = (\frac{n(n+1)}{2})^2 \qquad (8)$$

Alternatively, using the notation t_n introduced in section 2, formula (8) can be written as

$$N_{rects}[n] = (t_n)^2.$$

In particular, the problem of finding the number of rectangles on the checkerboard can be used as a motivation for the development of formula (2).

Formula (8) can be proved by the method of mathematical induction. To this end, one can note that the transition from $n \times n$ multiplication table to $(n+1) \times (n+1)$ multiplication table augments the former by the (n + 1)-st row and (n + 1)-st column the sum of numbers in which is equal to

$$2[(n+1)+2(n+1)+...+n(n+1)]+(n+1)^2$$
$$=2(n+1)(1+2+...+n)+(n+1)^2$$
$$=n(n+1)^2+(n+1)^2=(n+1)^3.$$

	A	B	C	D	E	F	G	H	I	J	K	L	M
1	←	→											
2	8	1	2	3	4	5	6	7	8				
3	1	1	2	3	4	5	6	7	8				
4	2	2	4	6	8	10	12	14	16				
5	3	3	6	9	12	15	18	21	24				
6	4	4	8	12	16	20	24	28	32				
7	5	5	10	15	20	25	30	35	40				
8	6	6	12	18	24	30	36	42	48				
9	7	7	14	21	28	35	42	49	56				
10	8	8	16	24	32	40	48	56	64				
11													
12													
13													
14													

Figure 1.5. The 8×8 multiplication table.

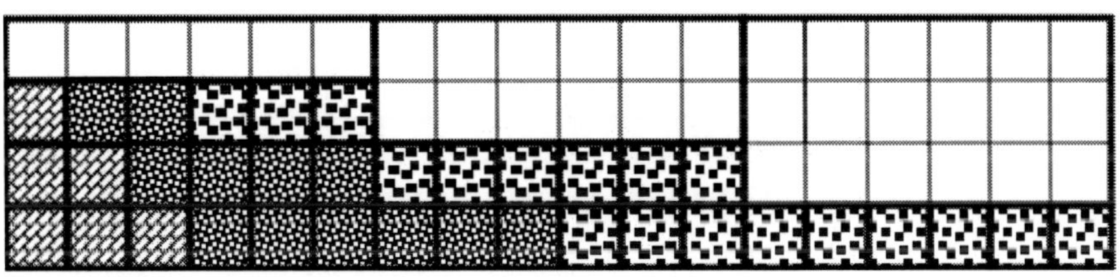

Figure 1.6. Twice the sum of the first three cubes of counting numbers.

Therefore, assuming formula (8) to be true implies that

$$N_{rects}[n+1]=N_{rects}[n]+(n+1)^3=[\frac{n(n+1)}{2}]^2+(n+1)^3$$
$$=\frac{(n+1)^2}{4}(n^2+4n+4)=[\frac{(n+1)(n+2)}{2}]^2.$$

This completes the proof of formula (8).

Remark 2. The proof of formula (8) was based on the following noteworthy property of numbers in the multiplication table: the sum of numbers that belong to the n-th row and n-th column of the table (a geometric structure sometimes referred to as *gnomon*) is equal to the cube of n. As the $n\times n$ multiplication table is the unity of n gnomons, the sum of the first n

cubes of counting numbers is equal to the sum of all numbers in the table. Two other connections between the table and the sums of perfect powers are: the sum of the first n counting numbers is equal to the sum of all numbers in the first row (or column) of the table and the sum of the first n squares of counting numbers is equal to the sum of all numbers that belong to the main (top left—bottom right) diagonal of the table.

Remark 3. The above-mentioned connection between the sum of cubes and the multiplication table can be used for conjecturing formula (5) through its geometric construction. Consider the case of $1^3 + 2^3 + 3^3$. Using the 3×3 multiplication table (see the corresponding fragment of a larger table pictured in Figure 1.5), one can represent this sum of three cubes as three sums of three entries of the table

$$1^3+2^3+3^3=(1+2+3)+(2+4+6)+(3+6+9).$$

Geometrically, the first, second, and third sums can be represented, respectively, through the shaded blocks in the third, second, and first rows of the diagram of Figure 1.6. Then, the shaded part (a ladder) can be augmented by three non-shaded rectangles with 6, 12, and 18 blocks to have a large rectangle with 72 blocks. As the number of shaded and non-shaded blocks is the same, we have $1^3+2^3+3^3=36$. Furthermore, the large rectangle has side lengths (3 + 1) and $3\cdot(1+2+3)$ that can be generalized to the case of n cubes as follows:

$$(3+1)\rightarrow(n+1),\ 3\cdot(1+2+3)\rightarrow n(1+2+3+...+n)=n\frac{n(n+1)}{2}.$$

From here one can conjecture formula (5).

5.1. Finding the Number of Squares on the $n\times n$ Checkerboard

Problem 1. How many squares are there on the $n\times n$ checkerboard?

Solution. The answer to this question immediately follows from the checkerboard problem if one recognizes that all squares of side l can be mapped to the cell of the multiplication table that corresponds to the product $(n-l+1)(n-l+1)$. This means that the total number of squares of side l is equal to $(n-l+1)(n-l+1)$. Indeed, considering as the basic square of side length l the one having the bottom-right vertex in the bottom-right corner of the $n\times n$ checkerboard (multiplication table), this basic square can be shifted up and to the left $n-1$ times thus, according to the rule of product (Chapter 5), making the total count of such squares equal to $(n-l+1)(n-l+1)$. When l changes from 1 to n, this product changes from n^2 to 1. Therefore, the total number of squares of size $n\times n, (n-1)\times(n-1)$, $(n-2)\times(n-2)$, ..., and 1×1 on the $n\times n$ checkerboard (multiplication table), for which the notation $N_{squares}[n]$ will be used, is equal to, respectively, 1^2, 2^2, 3^2, ..., and n^2. Using formula (4) yields

$$N_{squares}[n] = \frac{n(n+1)(2n+1)}{6} \qquad (9)$$

Remark 4. Ironically, a special case of the checkerboard problem—finding the number of squares on the checkerboard—requires the use of rather complicated summation machinery, namely, formula (4). On the other hand, the proof of formula (8) is more complicated in comparison with the proof of formula (4). This observation is instructive for it provides an example of how a more general problem may be easier to solve in comparison with a special case.

5.2. Finding the Number of Prisms within a Cube

Rubik's Cube[3] Problem. Find the total number of right rectangular prisms within the $n \times n \times n$ Rubik's cube ($n = 3$ in Figure 1.7).

Figure 1.7. The $3 \times 3 \times 3$ Rubik's cube.

Solution. Consider the cube as a combination of n identical prisms of the unit height, that is, n layers of the cube. Each such prism, consisting of $n \times n$ unit cubes, can be interpreted as the $n \times n$ multiplication table and a unit cube can be interpreted as a cell of the table. In that

[3] Rubik's cube (Figure 1.7) is a three-dimensional mechanical puzzle invented by a Hungarian architect Ernö Rubik in 1974 and nowadays is considered as one of the world's most famous toys.

way, there are n multiplication tables of size $n \times n$ within the $n \times n \times n$ cube. Each product in the multiplication table at layer l is equal to $i \times k \times l$ where i, k, = 1, 2, … , n. Every prism of height l, $1 \leq l \leq n$, has a base that belongs to one of the $n - l + 1$ layers of the $n \times n \times n$ cube. For example, within a $6 \times 6 \times 6$ cube, a base of a prism of two units in height can be a rectangle that belongs to five different layers (the cube's own base and four layers above it). Therefore, every prism of height l within the $n \times n \times n$ cube can be uniquely mapped on its base, a rectangle. One can count the number of prisms by counting the bases of the prisms, that is, by counting rectangles on a square grid (checkerboard). In turn, the number of such rectangles is the sum of all numbers in the corresponding multiplication table. Therefore, the number of prisms of height n is equal to $(t_n)^2$, the number of prisms of height $n - 1$ is equal to $2(t_n)^2$, the number of prisms of height $n - 2$ is equal to $3(t_n)^2$, ... , the number of prisms of height one is equal to $n(t_n)^2$. So, the total number of prisms within the $n \times n \times n$ cube (for which the notation $N_{prisms}[n]$ will be used) is equal to

$$
\begin{aligned}
&N_{prisms}[n] = (t_n)^2 + 2(t_n)^2 + 3(t_n)^2 + ... + n(t_n)^2 \\
&= (t_n)^2(1 + 2 + 3 + ... + n) = (t_n)^3
\end{aligned}
.
$$

Alternatively,

$$
N_{prisms}[n] = (\frac{n(n+1)}{2})^3 \tag{10}
$$

Problem 2. Find the total number of cubes within the $n \times n \times n$ cube.

Solution. Just like every prism of height l can be uniquely mapped onto its rectangular base, every cube of side l, $1 \leq l \leq n$, can be uniquely mapped on its square base. There are n^2 unit squares within the $n \times n$ square. Therefore, as there are n layers of unit cubes within the cube, there are $n \cdot (n^2) = n^3$ unit cubes within the cube. Next, there are $(n - 1)^2$ squares of side two within the $n \times n$ square. Because any cube of side two may reside within exactly two consecutive layers, there are $(n-1) \cdot (n-1)^2 = (n-1)^3$ such cubes. In general, there are $(n - l + 1)^3$ cubes of side l within the $n \times n \times n$ cube. So, the total number of cubes within the $n \times n \times n$ cube (for which the notation $S_{cubes}[n]$ will be used) is equal to the sum $1^3 + 2^3 + 3^3 + ... + n^3$. Using formula (5) yields

$$
S_{cubes}[n] = (\frac{n(n+1)}{2})^2 \tag{11}
$$

Note that the Rubik's cube problem can be used as a motivation for the development of formula (5). In that way, just like formulas (2)—(4) can be developed in the context of the checkerboard problem, formula (5) can be developed as one explores numerical properties of the Rubik's cube.

Remark 5. Comparing formulas (11) and (8) implies that the sum of all numbers in the $n \times n$ multiplication table is equal to $1^3 + 2^3 + 3^3 + ... + n^3$. The latter can be shown to represent the sum of all cubes within a $n \times n \times n$ cube. In other words, the number of rectangles within the $n \times n$ checkerboard is equal to the number of cubes within the $n \times n \times n$ cube. In order to explain this unexpected connection, all cubes in the $n \times n \times n$ cube can be arranged into n groups depending on the size of a cube. Then each such group can be mapped on a gnomon in one of the faces of the $n \times n \times n$ cube. By representing a face of the cube as the $n \times n$ multiplication table, and the technique used to prove formula (8), one can show that the sum of all numbers in the k-th gnomon is equal to k^3—the number of cubes in the corresponding group. The sum of numbers in all n gnomons is equal to the total number of rectangles within the $n \times n$ face of the $n \times n \times n$ cube. In that way, geometric structures of different dimensions (cubes and rectangles) can become connected through the appropriate use of the multiplication table.

5.3. Counting Rectangles with Special Properties on the Checkerboard

Problem 3. How many squares with a side measured by an odd number are there on the $(2n) \times (2n)$ checkerboard?

	A	B	C	D	E	F	G	H	I	J	K	L	M
1	←	→											
2	10	1	2	3	4	5	6	7	8	9	10		
3	1	1	2	3	4	5	6	7	8	9	10		
4	2	2	4	6	8	10	12	14	16	18	20		
5	3	3	6	9	12	15	18	21	24	27	30		
6	4	4	8	12	16	20	24	28	32	36	40		
7	5	5	10	15	20	25	30	35	40	45	50		
8	6	6	12	18	24	30	36	42	48	54	60		
9	7	7	14	21	28	35	42	49	56	63	70		
10	8	8	16	24	32	40	48	56	64	72	80		
11	9	9	18	27	36	45	54	63	72	81	90		
12	10	10	20	30	40	50	60	70	80	90	100		
13													
14													

Figure 1.8. Squares with odd side lengths on an even size board.

Solution. To begin, consider the case $n = 5$ and the 10×10 checkerboard (multiplication table) pictured in Figure 1.8. The squares with the side lengths 1, 3, 5, 7, and 9 can be put into one-to-one correspondence, respectively, with the square numbers 100, 64, 36, 16, and 4 located on the main diagonal of the table. Their sum is equal to 220.

In general, on the $(2n)\times(2n)$ checkerboard, all squares of side length $2l-1$ can be mapped onto the cell of the corresponding multiplication table that contains the product $(2n-(2l-1)+1)(2n-(2l-1)+1)=(2(n-l+1))^2$—a quantity that shows the total number of such squares; here $1 \le l \le n$, thereby, when $l = n$ we have $(2(n-l+1))^2=(2(n-n+1))^2=2^2$; when $l = 1$ we have

$$(2(n-l+1))^2=(2(n-1+1))^2=(2n)^2.$$

Using formula (4), the sum of the first n squares of even numbers can be found as follows:

$$2^2+4^2+6^2+...+(2n)^2=2^2(1^2+2^2+3^2+...+n^2)=\frac{2}{3}n(n+1)(2n+1).$$

Therefore, using the notation $N^{odd}_{squares}[2n]$ to represent the number of squares with side measured by an odd number, one can write

$$N^{odd}_{squares}[2n]=\frac{2}{3}n(n+1)(2n+1) \qquad (12)$$

In particular, when $n = 5$, formula (12) gives 220 squares, thereby, confirming the special case discussed above. Furthermore, the spreadsheet of Figure 1.8 can be modified to display and add the highlighted numbers only, thereby providing a setting for experimental verification of formula (12).

Problem 4. How many squares with a side measured by an odd number are there on the $(2n+1)\times(2n+1)$ checkerboard?

	A	B	C	D	E	F	G	H	I	J	K	L	M
1	←	→											
2	9	1	2	3	4	5	6	7	8	9			
3	1	1	2	3	4	5	6	7	8	9			
4	2	2	4	6	8	10	12	14	16	18			
5	3	3	6	9	12	15	18	21	24	27			
6	4	4	8	12	16	20	24	28	32	36			
7	5	5	10	15	20	25	30	35	40	45			
8	6	6	12	18	24	30	36	42	48	54			
9	7	7	14	21	28	35	42	49	56	63			
10	8	8	16	24	32	40	48	56	64	72			
11	9	9	18	27	36	45	54	63	72	81			
12													
13													
14													

Figure 1.9. Locating squares with odd side lengths on an odd size checkerboard.

Solution. To begin, consider the case $n = 4$ and the 9×9 checkerboard (multiplication table) pictured in Figure 1.9. The squares with side lengths 1, 3, 5, 7, and 9 can be put into one-to-one correspondence, respectively, with the square numbers 81, 49, 25, 9, and 1 located on the main diagonal of the table. Their sum is equal to 165.

In general, on the $(2n+1)\times(2n+1)$ checkerboard, all squares with the side length $2l - 1$ can be mapped on the cell of the corresponding multiplication table that includes the product $(2n+1-(2l+1)+1)(2n+1-(2l+1)+1) = (2(n-l)+3)^2$—a quantity that shows the total number of such squares; here $1 \le l \le n + 1$. When $l = n + 1$ and $l = 1$ one has, respectively, $(2(n-l)+3)^2 = (2(n-n-1)+3)^2 = 1$ and $(2(n-l)+3)^2 = (2(n-1)+3)^2 = (2n+1)^2$. Using formulas (4) and (12), the sum of the first n squares of odd numbers can be found as follows:

$$
\begin{aligned}
&1^2 + 3^2 + 5^2 + \ldots + (2n+1)^2 \\
&= 1^2 + 2^2 + 3^2 + 4^2 + \ldots + (2n)^2 + (2n+1)^2 - (2^2 + 4^2 + \ldots + (2n)^2) \\
&= \frac{(2n+1)(2n+2)(2(2n+1)+1)}{6} - \frac{2n(n+1)(2n+1)}{3} \\
&= \frac{(2n+1)(n+1)(4n+3)}{3} - \frac{2n(n+1)(2n+1)}{3} = \frac{(n+1)(4n^2+8n+3)}{3}.
\end{aligned}
$$

Finally, using the notation $N^{odd}_{squares}[2n+1]$ to represent the number of squares with a side measured by an odd number on the $(2n+1)\times(2n+1)$ checkerboard, one can write

$$N^{odd}_{squares}[2n+1]=\frac{(n+1)(4n^2+8n+3)}{3} \qquad (13)$$

Note that when n = 4, formula (13) gives 165 squares, thereby, confirming the special case shown in Figure 1.9. Furthermore, the spreadsheet can be modified to display and add the highlighted numbers only, thereby providing a setting for experimental verification of formula (13).

Problem 5. How many rectangles with at least one side length measured by an odd number can be found on the $2n \times 2n$ checkerboard?

Solution. One of the strategies to solve the problem is to find the number of rectangles both sides of which are measured by an even number and subtract this number from the total number of rectangles on the $2n \times 2n$ checkerboard. In other words, $N^{at\,least\,one\,odd}_{rects}[2n] = N_{rects}[2n] - N^{both\,even}_{rects}[2n]$, where the notations $N^{at\,least\,one\,odd}_{rects}[2n]$ and $N^{both\,even}_{rects}[2n]$ represent, respectively, the number of rectangles at least one side of which is measured by an odd number and the number of rectangles both sides of which are measured by an even number in the $2n \times 2n$ checkerboard.

To clarify, once again, consider the case n = 5. All rectangles with both sides measured by an even number can be put into one-to-one correspondence with the set of all odd numbers in the table (in Figure 1.10, these numbers are highlighted). For example, the number of 2×4 rectangles is equal to the product $9 \times 7 = (10-2+1) \times (10-4+1)$.

	A	B	C	D	E	F	G	H	I	J	K	L	M
1	← →												
2	10	1	2	3	4	5	6	7	8	9	10		
3	1	1	2	3	4	5	6	7	8	9	10		
4	2	2	4	6	8	10	12	14	16	18	20		
5	3	3	6	9	12	15	18	21	24	27	30		
6	4	4	8	12	16	20	24	28	32	36	40		
7	5	5	10	15	20	25	30	35	40	45	50		
8	6	6	12	18	24	30	36	42	48	54	60		
9	7	7	14	21	28	35	42	49	56	63	70		
10	8	8	16	24	32	40	48	56	64	72	80		
11	9	9	18	27	36	45	54	63	72	81	90		
12	10	10	20	30	40	50	60	70	80	90	100		
13													
14													

Figure 1.10. Rectangles with even dimensions on an even size checkerboard.

In general, the number of $2l \times 2k$ rectangles in the $2n \times 2n$ checkerboard, $1 \le k, l \le n$, is equal to the product $(2(n-l)+1) \times (2(n-k)+1)$. The sum of all such products, $N^{both\,even}_{rects}[2n]$, can be found as follows:

$$(1+3+...+2n-1)+3(1+3+...+2n-1)+...+(2n-1)(1+3+...+2n-1)$$
$$=(1+3+...+2n-1)(1+3+...+2n-1)=(1+3+...+2n-1)^2=(n^2)^2=n^4.$$

Substituting $2n$ for n in formula (8) yields

$$N_{rects}^{at\,least\,one\,odd}[2n]=(\frac{2n(2n+1)}{2})^2-n^4=n^2(2n+1)^2-n^4=n^2(n+1)(3n+1).$$

Therefore,

$$N_{rects}^{at\,least\,one\,odd}[2n]=n^2(n+1)(3n+1) \tag{14}$$

In particular, when n = 1, i.e., for the 2×2 checkerboard, formula (14) gives eight rectangles: four 1×1 rectangles, two 1×2 rectangles, and two 2×1 rectangles. The only 2×2 rectangle with both sides measured by an even number is the checkerboard itself. Furthermore, the spreadsheet of Figure 1.10 can be modified to display and add the non-highlighted (or highlighted) numbers only, thereby providing a setting for experimental verification of formula (14).

Problem 6. How many rectangles with at least one side measured by an odd number can be found on the $(2n+1)\times(2n+1)$ checkerboard?

Solution. Using an indirect way of counting $N_{rects}^{at\,least\,one\,odd}[2n+1]$ employed in the case of $2n\times 2n$ checkerboard, Figure 1.11 (where n = 4) and formula (2), one can proceed as follows

$$N_{rects}^{both\,even}[2n+1]=(4+8+12+...+4n)+2(4+8+12+...+4n)$$
$$+3(4+8+12+...+4n)+...+n(4+8+12+...+4n)$$
$$=(4+8+12+...+4n)(1+2+3+...+n)=4(1+2+3+...+n)^2=n^2(n+1)^2$$

and then find

$$N_{rects}^{at\,least\,one\,odd}[2n+1]=N_{rects}[2n+1]-N_{rects}^{both\,even}[2n+1]$$
$$=(2n+1)^2(n+1)^2-n^2(n+1)^2$$
$$=(n+1)^2(2n+1-n)(2n+1+n)=(n+1)^3(3n+1).$$

Therefore,

$$N_{rects}^{at\,least\,one\,odd}[2n+1]=(n+1)^3(3n+1) \tag{15}$$

In particular, when n = 1, i.e., for the 3×3 checkerboard, formula (15) gives 32 rectangles: nine 1×1 rectangles (squares), six 1×2 rectangles, three 1×3 rectangles, six

2×1 rectangles, three 3×1 rectangles, two 2×3, two 3×2, and one 3×3 rectangle (square). The only type of a rectangle on the 3×3 checkerboard with both sides measured by an even number is a 2×2 square. The number of such rectangles (squares) is equal to $N_{rects}^{botheven}[3]=4$. Furthermore, the spreadsheet of Figure 1.11 can be modified to display and add the non-highlighted (or highlighted) numbers only, thereby providing a setting for experimental verification of formula (15).

Problem 7. How many rectangles with both sides measured by an odd number are there on the $2n\times 2n$ checkerboard?

Solution. Let $N_{rects}^{bothodd}[2n]$ represent the number of rectangles sought. Then (see Figure 1.12 where $n = 5$)

$$N_{rects}^{bothodd}[2n]=(4+8+12+...+4n)+2(4+8+12+...+4n)$$
$$+3(4+8+12+...+4n)+...+n(4+8+12+...+4n)=4(1+2+3+...+n)^2$$

Using formula (2) yields

$$N_{rects}^{bothodd}[2n]=n^2(n+1)^2 \qquad (16)$$

	A	B	C	D	E	F	G	H	I	J	K	L	M
1	←	→											
2	9	1	2	3	4	5	6	7	8	9			
3	1	1	2	3	4	5	6	7	8	9			
4	2	2	4	6	8	10	12	14	16	18			
5	3	3	6	9	12	15	18	21	24	27			
6	4	4	8	12	16	20	24	28	32	36			
7	5	5	10	15	20	25	30	35	40	45			
8	6	6	12	18	24	30	36	42	48	54			
9	7	7	14	21	28	35	42	49	56	63			
10	8	8	16	24	32	40	48	56	64	72			
11	9	9	18	27	36	45	54	63	72	81			
12													
13													
14													

Figure 1.11. Rectangles with even dimensions on an odd size checkerboard.

The spreadsheet of Figure 1.12 can be modified to display and add the highlighted numbers only, thereby providing a setting for experimental verification of formula (16).

	A	B	C	D	E	F	G	H	I	J	K	L	M
1	←	→											
2	10	1	2	3	4	5	6	7	8	9	10		
3	1	1	2	3	4	5	6	7	8	9	10		
4	2	2	4	6	8	10	12	14	16	18	20		
5	3	3	6	9	12	15	18	21	24	27	30		
6	4	4	8	12	16	20	24	28	32	36	40		
7	5	5	10	15	20	25	30	35	40	45	50		
8	6	6	12	18	24	30	36	42	48	54	60		
9	7	7	14	21	28	35	42	49	56	63	70		
10	8	8	16	24	32	40	48	56	64	72	80		
11	9	9	18	27	36	45	54	63	72	81	90		
12	10	10	20	30	40	50	60	70	80	90	100		
13													
14													

Figure 1.12. Rectangles with odd dimensions on an even size checkerboard.

Problem 8. How many rectangles with both sides measured by an odd number are there on the $(2n+1)\times(2n+1)$ checkerboard?

Solution. Let $N_{rects}^{bothodd}[2n+1]$ represent the number of rectangles sought. Then (see Figure 1.13 where $n = 4$)

$$N_{rects}^{both\,odd}[2n+1] = (1+3+5+...+2n+1)$$
$$+3(1+3+5+...+2n+1)$$
$$+5(1+3+5+...+2n+1)+...+(2n+1)(1+3+5+...+2n+1)$$
$$=(1+3+5+...+2n+1)^2.$$

Using formula (3) yields

$$N_{rects}^{bothodd}[2n+1] = (n+1)^4 \tag{17}$$

The spreadsheet of Figure 1.13 can be modified to display and add the highlighted numbers only, thereby providing a setting for experimental verification of formula (17).

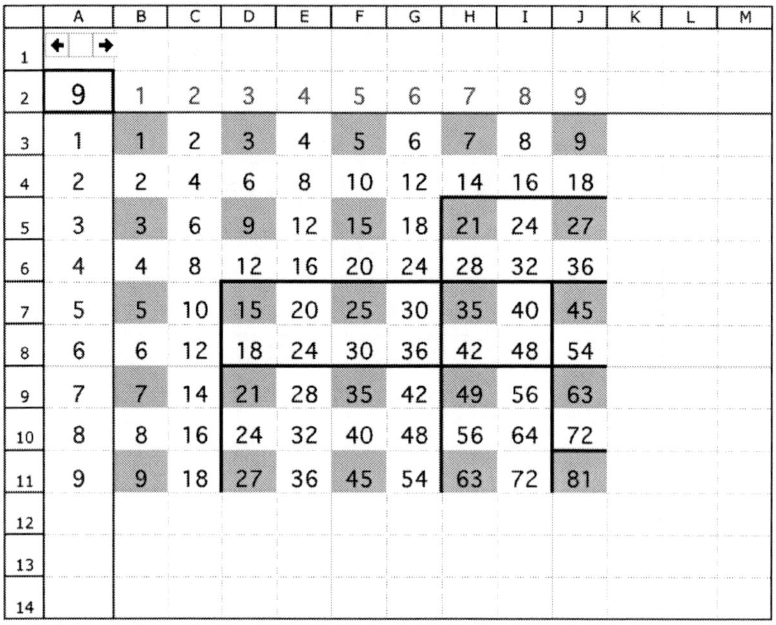

	A	B	C	D	E	F	G	H	I	J	K	L	M
1													
2	9	1	2	3	4	5	6	7	8	9			
3	1	1	2	3	4	5	6	7	8	9			
4	2	2	4	6	8	10	12	14	16	18			
5	3	3	6	9	12	15	18	21	24	27			
6	4	4	8	12	16	20	24	28	32	36			
7	5	5	10	15	20	25	30	35	40	45			
8	6	6	12	18	24	30	36	42	48	54			
9	7	7	14	21	28	35	42	49	56	63			
10	8	8	16	24	32	40	48	56	64	72			
11	9	9	18	27	36	45	54	63	72	81			
12													
13													
14													

Figure 1.13. Rectangles with odd dimensions on an odd size board.

6. Comparing the Rates of Growth of Different Families of Geometric Figures

Formulas developed in this chapter for counting the number of geometric figures within the $n \times n$ checkerboard and $n \times n \times n$ Rubik's cube demonstrated that such a number grows larger as n increases. However, the rates of growth of different families of geometric figures may be different. Different rates of growth of functions were used in section 3 as a means of explaining a counterexample to an inductively generated conjecture. Exploring the rates of growth of different functions requires the integration of tools of algebra, discrete mathematics, and calculus.

6.1. Comparing Different Sets of Rectangles on the $n \times n$ Checkerboard

Proposition 4. Let $r(n) = \dfrac{N_{rects}^{at\,least\,one\,odd}[2n]}{N_{rects}[2n]}$ —the ratio of the number of rectangles with at least one side measured by an odd number to the total number of rectangles on the $2n \times 2n$ checkerboard. Then

$$\lim_{n \to \infty} r(n) = \frac{3}{4}.$$

Proof. It follows from formulas (8) and (14) that

$$\lim_{n\to\infty} r(n) = \lim_{n\to\infty} \frac{N_{rects}^{at\,least\,one\,odd}[2n]}{N_{rects}[2n]} = \lim_{n\to\infty} \frac{n^2(n+1)(3n+1)}{n^2(2n+1)^2}$$

$$= \frac{3}{4} \lim_{n\to\infty} \frac{(1+\frac{1}{n})(1+\frac{1}{3n})}{(1+\frac{1}{2n})^2} = \frac{3}{4}.$$

Proposition 5. Let $r(n) = \dfrac{N_{rects}^{at\,least\,one\,odd}[2n+1]}{N_{rects}[2n+1]}$—the ratio of the number of rectangles with at least one side measured by an odd number to the total number of rectangles on the $(2n+1)\times(2n+1)$ checkerboard. Then

$$\lim_{n\to\infty} r(n) = \frac{3}{4}.$$

Proof. It follows from formulas (8) and (15) that

$$\lim_{n\to\infty} r(n) = \lim_{n\to\infty} \frac{N_{rects}^{at\,least\,one\,odd}[2n+1]}{N_{rects}[2n+1]} = \lim_{n\to\infty} \frac{(n+1)^3(3n+1)}{(n+1)^2(2n+1)^2}$$

$$= \frac{3}{4} \lim_{n\to\infty} \frac{(1+\frac{1}{n})(1+\frac{1}{3n})}{(1+\frac{1}{2n})^2} = \frac{3}{4}.$$

Proposition 6. Let $r(n) = \dfrac{N_{rects}^{both\,odd}[2n]}{N_{rects}[2n]}$—the ratio of the number of rectangles with both sides measured by an odd number to the total number of rectangles on the $2n\times 2n$ checkerboard. Then

$$\lim_{n\to\infty} r(n) = \frac{1}{4}.$$

Proof. It follows from formulas (8) and (16) that

$$\lim_{n\to\infty} r(n) = \lim_{n\to\infty} \frac{N_{rects}^{both\,odd}[2n]}{N_{rects}[2n]} = \lim_{n\to\infty} \frac{n^2(n+1)^2}{n^2(2n+1)^2} = \frac{1}{4} \lim_{n\to\infty} \frac{(1+\frac{1}{n})^2}{(1+\frac{1}{2n})^2} = \frac{1}{4}.$$

Proposition 7. Let $r(n)=\dfrac{N_{rects}^{both\,odd}[2n+1]}{N_{rects}[2n+1]}$—the ratio of the number of rectangles with both sides measured by an odd number to the total number of rectangles on the $(2n+1)\times(2n+1)$ checkerboard. Then

$$\lim_{n\to\infty} r(n)=\frac{1}{4}.$$

Proof. It follows from formulas (8) and (17) that

$$\lim_{n\to\infty} r(n)=\lim_{n\to\infty}\frac{N_{rects}^{both\,odd}[2n+1]}{N_{rects}[2n+1]}=\lim_{n\to\infty}\frac{(n+1)^4}{(2n+1)^2(n+1)^2}$$

$$=\frac{1}{4}\lim_{n\to\infty}\frac{(1+\frac{1}{n})^4}{(1+\frac{1}{2n})^2(1+\frac{1}{n})^2}=\frac{1}{4}.$$

6.2. Comparing the Number of Squares to the Number of Rectangles on the $n\times n$ Checkerboard

Proposition 8. Let $r(n)=\dfrac{N_{squares}[n]}{N_{rects}[n]}$—the ratio of the number of squares to the number of rectangles on the $n\times n$ checkerboard. Then

$$\lim_{n\to\infty} r(n)=0 \tag{18}$$

Proof. It follows from formulas (8) and (9) that

$$\lim_{n\to\infty} r(n)=\lim_{n\to\infty}\frac{N_{squares}[n]}{N_{rects}[n]}=\lim_{n\to\infty}\frac{\frac{n(n+1)(2n+1)}{6}}{\frac{n^2(n+1)^2}{4}}$$

$$=\frac{2}{3}\lim_{n\to\infty}\frac{2n+1}{n(n+1)}=\frac{2}{3}\lim_{n\to\infty}\frac{\frac{2}{n}+\frac{1}{n^2}}{1+\frac{1}{n}}=0$$

This completes the proof.

The limiting behavior of the function $r(n)$ expressed through relation (18) indicates that although both the number of rectangles and the number of squares on an $n\times n$ checkerboard increases as n grows large, the growth of rectangles is much faster than the growth of squares. In other words, for any positive number ε, however small it is, there exists an $N\times N$

checkerboard where the fraction of squares among rectangles is smaller than ε. There are problems in mathematics, in particular, those arising in the context of deciding the convergence of series, where the evaluation of the order of "smallness" of variable quantities is important. The following proposition evaluates the "smallness" of the ratio $r(n)$ by squeezing it between infinitely small quantities expressed in terms of the size of a checkerboard.

Proposition 9. Let $r(n)$ be the ratio of the number of squares to the number of rectangles on the $n \times n$ checkerboard. Then the inequalities

$$\frac{1}{n} \leq r(n) \leq \frac{1}{\sqrt{n}} \tag{19}$$

hold true for $n = 1, 2, 3, \ldots$.

Proof. As was already shown above, formulas (8) and (9) yield

$$r(n) = \frac{2(2n+1)}{3n(n+1)}.$$

Evaluating the difference between $r(n)$ and $\frac{1}{n}$ yields

$$r(n) - \frac{1}{n} = \frac{2(2n+1)}{3n(n+1)} - \frac{1}{n} = \frac{4n+2-3(n+1)}{3n(n+1)} - \frac{n-1}{3n(n+1)} \geq 0 \text{ for } n \geq 1.$$

Therefore, $r(n) \geq \frac{1}{n}$ and $r(n) = \frac{1}{n}$ when $n = 1$.

Setting $\sqrt{n} = x$, the difference between $r(n)$ and $\frac{1}{\sqrt{n}}$ can be evaluated as follows

$$r(n) - \frac{1}{\sqrt{n}} = \frac{2(2n+1)}{3n(n+1)} - \frac{1}{\sqrt{n}} = \frac{2(2x^2+1)}{3(x^4+x^2)} - \frac{1}{x} = \frac{x^2+2-3x^3}{3x^2(x^2+1)}$$
$$= -\frac{3x^3 - x^2 - 2}{3x^2(x^2+1)} = -\frac{(x-1)(3x^2+2x+2)}{3x^2(x^2+1)} \leq 0$$

for all $x = \sqrt{n} \geq 1$.

Therefore, $r(n) \leq \frac{1}{\sqrt{n}}$ and $r(n) = \frac{1}{\sqrt{n}}$ when $n = 1$. This completes the proof.

6.3. Harmonic Series and the Method of d'Oresme

Formula (2) implies that the sum of the first n counting numbers tends to infinity as $n \to \infty$. How does the sum of the reciprocals of counting numbers, $1+\frac{1}{2}+\frac{1}{3}+...+\frac{1}{n}$, behave as $n \to \infty$? Unfortunately, there is no closed formula for the latter sum to evaluate this limit. Therefore, in order to address the above question, one moves from finite sums to the exploration of infinite sums called series. In particular, the infinite sum of the reciprocals of counting numbers, for which the notation $\sum_{n=1}^{\infty}\frac{1}{n}$ will be used below, is called the harmonic series.

Note that each term of this series beginning from the second is equal to the reciprocal of the arithmetic mean of the reciprocals of the two neighboring terms. Indeed, for all $n \geq 2$ we have the identity

$$\frac{1}{n}=\frac{1}{\frac{(n-1)+(n+1)}{2}}$$

where $n - 1$ and $n + 1$ are, respectively, the reciprocals of $\frac{1}{n-1}$ and $\frac{1}{n+1}$. For example, 1, $\frac{1}{2}$, and $\frac{1}{3}$ are the first three terms of the harmonic series and the second term $\frac{1}{2}=\frac{1}{\frac{1+3}{2}}$. In turn, the reciprocal of the average of two numbers is called the harmonic mean of these numbers. This property of the reciprocals of counting numbers explains the term *harmonic series*. In general, the fraction $\frac{n}{\frac{1}{a_1}+\frac{1}{a_2}+...+\frac{1}{a_n}}$ is the harmonic mean of the numbers $a_i > 0,\ i = 1, 2, ..., n$.

Proposition 10. The harmonic series $\sum_{n=1}^{\infty}\frac{1}{n}$ diverges. In other words, the infinite sum of (monotonically decreasing) reciprocals of counting numbers is infinite.

Proof. The divergence of the harmonic series was originally proved by the 14th century French scholar Nicholas d'Oresme who suggested the method of grouping the terms of the series into the sums with the number of terms equal to a power of two, and then evaluating from below the sum of the terms in each group as follows

$$\sum_{n=1}^{\infty}\frac{1}{n}=1+(\frac{1}{2})+\underbrace{(\frac{1}{3}+\frac{1}{4})}_{two\ terms}+\underbrace{(\frac{1}{5}+\frac{1}{6}+\frac{1}{7}+\frac{1}{8})}_{four\ terms}+\underbrace{(\frac{1}{9}+\frac{1}{10}+...+\frac{1}{16})}_{eight\ terms}+\underbrace{(\frac{1}{17}+\frac{1}{18}+...+\frac{1}{32})}_{sixteen\ terms}+...$$

$$>1+(\frac{1}{2})+\underbrace{(\frac{1}{4}+\frac{1}{4})}_{two\ terms}+\underbrace{(\frac{1}{8}+\frac{1}{8}+\frac{1}{8}+\frac{1}{8})}_{four\ terms}+\underbrace{(\frac{1}{16}+\frac{1}{16}+...+\frac{1}{16})}_{eight\ terms}+\underbrace{(\frac{1}{32}+\frac{1}{32}+...+\frac{1}{32})}_{sixteen\ terms}+...$$

$$=1+\frac{1}{2}+2\cdot\frac{1}{4}+4\cdot\frac{1}{8}+8\cdot\frac{1}{16}+16\cdot\frac{1}{32}+...=1+\frac{1}{2}+\frac{1}{2}+...=+\infty$$

Remark 6. The divergence of the harmonic series demonstrates that although $\lim_{n\to\infty}\frac{1}{n}=0$, the reciprocals of counting numbers do not become "small enough" in order, as an infinite sum, to form a finite number. In that way, the relation $\lim_{n\to\infty} a(n)=0$ does not imply the convergence of the series $\sum_{n=1}^{\infty} a(n)$, but rather, the tendency of the sequence $a(n)$ to zero as n increases is only a necessary condition for the series to converge.

The harmonic series can be used as a tool for proving the convergence of other series.

Proposition 11. The series $\sum_{n=1}^{\infty} r(n)$ diverges. In other words, although the fraction of squares among rectangles on a square size checkerboard tends to zero as the size of the checkerboard increases, the infinite sum of such fractions is infinite as well (that is bigger than any given number).

Proof. According to (19), $r(n)\geq\frac{1}{n}$ and, thereby, due to Proposition 10, we have

$$\sum_{n=1}^{\infty} r(n)\geq\sum_{n=1}^{\infty}\frac{1}{n}=+\infty .$$

This completes the proof.

6.4. Comparing the Number of Cubes to the Number of Prisms in the Rubik's Cube

Proposition 12. Let $q(n)=\frac{N_{cubes}[n]}{N_{prisms}[n]}$ —the ratio of the number of cubes to the number of prisms within the $n\times n\times n$ Rubik's cube. Then

$$\lim_{n\to\infty} q(n)=0 \tag{20}$$

Proof. It follows from formulas (10) and (11) that

$$\lim_{n\to\infty} q(n)=\lim_{n\to\infty}\frac{(\frac{n(n+1)}{2})^2}{(\frac{n(n+1)}{2})^3}=\lim_{n\to\infty}\frac{2}{n(n+1)}=\lim_{n\to\infty}\frac{2}{n^2(1+\frac{1}{n})}=0\,.$$

This completes the proof.

Once again, it is interesting to evaluate the "smallness" of the ratio $q(n)$.

Proposition 13. The ratio $q(n)$ of the number of cubes to the number of prisms within the $n\times n\times n$ Rubik's cube satisfies the inequalities

$$\frac{1}{n^2}\le q(n)<\frac{2}{n^2} \tag{21}$$

for all n= 1, 2, 3, … .

Proof. To prove the inequality $q(n)\ge\frac{1}{n^2}$, one can evaluate the difference between $q(n)$ and $\frac{1}{n^2}$ as follows:

$$q(n)-\frac{1}{n^2}=\frac{2}{n(n+1)}-\frac{1}{n^2}=\frac{2n-(n+1)}{n^2(n+1)}=\frac{n-1}{n^2(n+1)}\ge 0$$

for all $n\ge 1$. Therefore $q(n)\ge\frac{1}{n^2}$ and $q(n)=\frac{1}{n^2}$ when $n=1$.

Evaluating the difference between $\frac{2}{n^2}$ and $q(n)$ yields

$$\frac{2}{n^2}-q(n)=\frac{2}{n^2}-\frac{2}{n(n+1)}=\frac{2(n+1-n)}{n^2(n+1)}=\frac{2}{n^2(n+1)}>0$$ for all $n\ge 1$. Therefore, $q(n)<\frac{2}{n^2}$ for n = 1, 2, 3, … .

This completes the proof of inequalities (21).

Remark 7. Another way to prove the inequality $q(n) < \frac{2}{n^2}$ is to note that $n^2 < n(n+1)$ whence $\frac{1}{n^2} > \frac{1}{n(n+1)}$. The last inequality makes it possible to write $\frac{2}{n^2} - \frac{2}{n(n+1)} > \frac{2}{n(n+1)} - \frac{2}{n(n+1)} = 0$, thereby confirming that $\frac{2}{n^2} > \frac{2}{n(n+1)} = q(n)$.

One may wonder as to why the inequality $n^2 < n(n+1)$ can be used in proving the inequality $q(n) < \frac{2}{n^2}$, yet it does not work when proving the inequality $q(n) \geq \frac{1}{n^2}$? To answer this question note that when comparing the three quantities $\frac{n^2}{2}$, $\frac{n(n+1)}{2}$ and n^2 for $n \geq 1$ the following observation can be made: $\frac{n^2}{2}$ and $\frac{n(n+1)}{2}$ never coincide, whereas $\frac{n(n+1)}{2}$ and n^2 do coincide when $n = 1$. That is why, in general, when proving an inequality between two quantities depending on n for $n \geq 1$, one cannot use a strict inequality between their components as this strict inequality does not hold true for $n = 1$. On the other hand, when one proves a non-strict inequality, establishing that type of inequality between its appropriate components can work at a means of proving. As Poincaré[4] noted, "[students] wish to know not merely whether all the syllogisms of a demonstration are correct, but why they link together in this order rather than another. In so far as to them they seem engendered by caprice and not by an intelligence always conscious of the end to be attained, they do not believe they understand" (cited in [Hadamard[5], 1996, p. 104]).

6.5. Converging Series

Inequalities (21) show that, similar to the sequence $r(n)$, the sequence $q(n)$ has been squeezed between two sequences that tend to zero as n grows large. Therefore, according to the Pinching Principle (Krantz, 2009), $\lim_{n\to\infty} q(n) = 0$. The last relation confirms the conclusion of Proposition 12. That is, the necessary condition for the convergence of the series $\sum_{n=1}^{\infty} q(n)$ is satisfied. However, unlike $r(n)$, the sequence $q(n)$ vanishes, as n increases, with such a rate that one may suspect the series $\sum_{n=1}^{\infty} q(n)$ to be bounded from above. One may recall that there exist infinite sums of infinitely small quantities that are finite. For example,

---

4 Henri Poincaré (1854 – 1912)—the outstanding French mathematician and physicist known for his major contributions to practically all branches of mathematics.

5 Jacques Salomon Hadamard (1865-1963)—a French mathematician who made important contributions to many branches of mathematics and, in particular, studied mathematical thinking process.

the sum of an infinitely decreasing geometric series is a finite number. Indeed, when $0 < s < 1$, the sum

$$1+s+s^2+...+s^n+...=\frac{1}{1-s} \qquad (22)$$

This note about converging geometric series allows for a modification of the method of d'Oresme that was used to prove the divergence of the harmonic series, to be utilized for the demonstration of

Proposition 14. The series $\sum_{n=1}^{\infty}\frac{1}{n^2}$ converges. In other words, an infinite sum of the reciprocals of squares of consecutive counting numbers is a finite number.

Proof. The terms of the series $\sum_{n=1}^{\infty}\frac{1}{n^2}$ can be arranged in groups so that the number of terms in each group is equal to a power of two. Then the sum of the terms in each group can be evaluated from above as follows

$$\sum_{n=1}^{\infty}\frac{1}{n^2}=1+(\frac{1}{2^2})+\underbrace{(\frac{1}{3^2}+\frac{1}{4^2})}_{\text{two terms}}+\underbrace{(\frac{1}{5^2}+\frac{1}{6^2}+\frac{1}{7^2}+\frac{1}{8^2})}_{\text{four terms}}+\underbrace{(\frac{1}{9^2}+\frac{1}{10^2}+...+\frac{1}{16^2})}_{\text{eight terms}}$$

$$+\underbrace{(\frac{1}{17^2}+\frac{1}{18^2}+...+\frac{1}{32^2})}_{\text{sixteen terms}}+...$$

$$<1+(\frac{1}{2^2})+\underbrace{(\frac{1}{2^2}+\frac{1}{2^2})}_{\text{two terms}}+\underbrace{(\frac{1}{4^2}+\frac{1}{4^2}+\frac{1}{4^2}+\frac{1}{4^2})}_{\text{four terms}}+\underbrace{(\frac{1}{8^2}+\frac{1}{8^2}+...+\frac{1}{8^2})}_{\text{eight terms}}+...$$

$$=1+\frac{1}{2^2}+2\cdot\frac{1}{2^2}+4\cdot\frac{1}{4^2}+8\cdot\frac{1}{8^2}+...+n\cdot\frac{1}{n^2}+...=\frac{1}{2^2}+1+\frac{1}{2}+\frac{1}{4}+\frac{1}{8}+...$$

$$=\frac{1}{4}+\frac{1}{1-\frac{1}{2}}=2\frac{1}{4}.$$

Here, the summation formula (22) was used in the case $s = 1/2$. Therefore, the infinite sum $\sum_{n=1}^{\infty}\frac{1}{n^2}$ is a finite number (the exact value of which was not known until the 18th century), that is, the series converges.

Remark 8. Euler[6] discovered the following remarkable formula

---

[6] Leonhard Euler (1707-1783)—the great Swiss mathematician, the father of all modern mathematics.

$$\sum_{n=1}^{\infty}\frac{1}{n^2}=\frac{\pi^2}{6} \quad (23)$$

and asserted that "six times the sum of this series is equal to the square of the circumference of a circle whose diameter is 1" (cited in Dunham, 1999, pp. 45-46, where the proof of formula (23) and its fascinating history can be found). One can use a spreadsheet (Figure 1.14) to compute the sum of the first 10,000 terms of the series and see that $|\sum_{n=1}^{10,000}\frac{1}{n^2}-\frac{\pi^2}{6}|=9.9\times10^{-5}$ —an equality that demonstrates the closeness of the two values. At the same time, in the case of diverging harmonic series, as the use of a spreadsheet can show, $\sum_{n=1}^{10,000}\frac{1}{n}=9.78...$, demonstrating an extremely slow divergence which can even be taken for convergence if approached from a computational perspective alone. More information on the divergence of the harmonic series can be found in Chapter 6.

Proposition 15. The series $\sum_{n=1}^{\infty}q(n)$ converges to the number 2.

Proof. Due to (21), $q(n)<\frac{2}{n^2}$ and, thereby, due to Proposition 14, we have $\sum_{n=1}^{\infty}q(n)<2\sum_{n=1}^{\infty}\frac{1}{n^2}<+\infty$.

However, noting that

$$q(n)=\frac{1}{t_n}=\frac{2}{n(n+1)}=2(\frac{1}{n}-\frac{1}{n+1})$$

◇	A	B	C	D
1	n	1/n^2	The sum	|The sum - n^2/6|
2	1	1	1	0.644934067
3	2	0.25	1.25	0.394934067
4	3	0.11111111	1.361111111	0.283822956
5	4	0.0625	1.423611111	0.221322956
6	5	0.04	1.463611111	0.181322956
7	6	0.02777778	1.491388889	0.153545178
8	7	0.02040816	1.511797052	0.133137015
9	8	0.015625	1.527422052	0.117512015
10	9	0.01234568	1.539767731	0.105166336
11	10	0.01	1.549767731	0.095166336
10001	10000	0.00000001	1.644834072	9.9995E-05

Figure 1.14. Numerical demonstration of the identity $\sum_{n=1}^{\infty}\frac{1}{n^2}=\frac{\pi^2}{6}$.

(cf. formula (6), Chapter 3), one can write

$$\sum_{n=1}^{\infty} q(n) = 2\sum_{n=1}^{\infty}(\frac{1}{n} - \frac{1}{n+1}) = 2[(1-\frac{1}{2}) + (\frac{1}{2} - \frac{1}{3}) + (\frac{1}{3} - \frac{1}{4}) + ...$$
$$+(\frac{1}{n-1} - \frac{1}{n}) + (\frac{1}{n} - \frac{1}{n+1}) + ...] = 2$$

Indeed, the infinite sum in the square brackets can be reduced to the number 1.

Remark 9. Besides the case of geometric series, such easiness of finding the exact sum of a series (in the case of Proposition 15, the infinite sum of the reciprocals of consecutive triangular numbers) through simple algebraic manipulations is a rare occurrence. Indeed, this was not the case for the summation of the reciprocals of square numbers (although, ironically, the expression n^2 is less complicated when compared to $\frac{n(n+1)}{2}$) and the discovery of formula (23), which escaped the efforts of many outstanding mathematicians, was considered a mathematical triumph of the 18th century.

6.6. Comparing the Rate of Growth of Squares and Rectangles with Special Properties

Proposition 16. Consider the $2n \times 2n$ checkerboard. Let

$u(n) = \frac{N_{squares}^{odd}[2n]}{N_{rects}^{at\ least\ one\ odd}[2n]}$ — the ratio of the number of odd side squares to the number of rectangles at least one side of which is measured by an odd number. Then

$$\lim_{n\to\infty} u(n) = 0 \qquad (24)$$

In other words, the order of magnitude of odd-sided squares is less that the order of magnitude of rectangles at least one side of which is measured by an odd number.

Proof. According to formulas (12) and (14) we have

$$u(n) = \frac{2n(n+1)(2n+1)}{3n^2(n+1)(3n+1)} = \frac{2(2n+1)}{3n(3n+1)}.$$

Therefore,

$$\lim_{n\to\infty} u(n) = \lim_{n\to\infty} \frac{2(2n+1)}{3n(3n+1)} = \frac{2}{3}\lim_{n\to\infty} \frac{2+\frac{1}{n}}{n(1+\frac{1}{n})} = 0 .$$

This proves relation (24).

In order to evaluate the "smallness" of $u(n)$, one can use a spreadsheet to investigate its behavior numerically prior to formal demonstration. To this end, one can introduce the function $f(n)=\frac{2n+1}{3n+1}$, $u(n)=\frac{2}{3n}f(n)$, and then model $f(n)$ numerically within a spreadsheet. As Figure 1.15 shows, $f(n) \le 3/4$ and it appears that $f(n) > 2/3 = 0.666...$. In that way, numerical modeling within a spreadsheet can motivate

Proposition 17. On the $2n \times 2n$ checkerboard, the sequence $u(n)=\frac{N_{squares}^{odd}[2n]}{N_{rects}^{at\,least\,one\,odd}[2n]}$ satisfies the inequalities

$$\frac{4}{9n} < u(n) \le \frac{1}{2n} \tag{25}$$

for all $n = 1, 2, 3, \dots$.

Proof. Numerical evidence can be confirmed by formal demonstration that $f(n)$ is a monotonically decreasing function. Indeed, the derivative

$$\frac{df(n)}{dn}=\frac{2(3n+1)-3(2n+1)}{(3n+1)^2}=-\frac{1}{(3n+1)^2}<0.$$

Therefore for $n \ge 1$ we have $f(n) \le f(1)=\frac{3}{4}$. Furthermore,

$$f(n)-\frac{2}{3}=\frac{2n+1}{3n+1}-\frac{2}{3}=\frac{3(2n+1)-2(3n+1)}{3(3n+1)}=\frac{1}{3(3n+1)}>0.$$

Therefore, $\frac{2}{3}<f(n)\le\frac{3}{4}$. Dividing both sides of the last relation by $1.5n$ yields (25). This completes the proof.

Proposition 18. The series $\sum_{n=1}^{\infty} u(n)$ diverges.

Proof. According to (25), $u(n) > \frac{4}{9n}$ and, thereby, due to Proposition 10, we have

$$\sum_{n=1}^{\infty} u(n) > \frac{4}{9}\sum_{n=1}^{\infty}\frac{1}{n} = +\infty .$$

	A	B
1	n	f(n)
2	1	0.75
3	2	0.714285714
4	3	0.7
5	4	0.692307692
6	5	0.6875
7	6	0.684210526
8	7	0.681818182
9	8	0.68
319	318	0.667015707
320	319	0.667014614

Figure 1.15. Motivating Proposition 17 through numerical evidence.

$$\frac{d}{dn}\frac{4\cdot n^2+8\cdot n+3}{3\cdot(n+1)\cdot(3\cdot n+1)}$$

$$\frac{1}{3}\frac{8n+8}{(n+1)(3n+1)}-\frac{1}{3}\frac{4n^2+8n+3}{(n+1)^2(3n+1)}-\frac{4n^2+8n+3}{(n+1)(3n+1)^2}$$

simplify(%)

$$-\frac{2}{3}\frac{4n^2+5n+2}{(n+1)^2(3n+1)^2}$$

Figure 1.16. Finding derivative using *Maple*.

This completes the proof.

Proposition 19. Consider the $(2n+1)\times(2n+1)$ checkerboard. Let $v(n)=\dfrac{N_{squares}^{odd}[2n+1]}{N_{rects}^{at\,least\,one\,odd}[2n+1]}$—the ratio of the number of odd-sided squares to the number of rectangles with at least one side measured by an odd number on this checkerboard. Then

$$\lim_{n\to\infty} v(n)=0$$

Proof. According to formulas (13) and (15),

$$v(n)=\frac{(n+1)(4n^2+8n+3)}{3(n+1)^3(3n+1)}=\frac{4n^2+8n+3}{3(n+1)^2(3n+1)}.$$

Therefore,

$$\lim_{n\to\infty} v(n)=\lim_{n\to\infty}\frac{4n^2+8n+3}{3(n+1)^2(3n+1)}=\frac{1}{3}\lim_{n\to\infty}\frac{n^2(4+\frac{8}{n}+\frac{3}{n^2})}{n^3(1+\frac{1}{n})^2(3+\frac{1}{n})}=0.$$

Proposition 20. On an odd size checkerboard, the sequence $v(n)$satisfies the inequalities

$$\frac{2}{9n} < v(n) < \frac{5}{8n} \tag{26}$$

for all $n = 1, 2, 3, \ldots$.

Proof. Consider the function

$$f(n) = \frac{4n^2 + 8n + 3}{3(n+1)(3n+1)}.$$

As the differentiation of $f(n)$ becomes cumbersome[7], one can use *Maple*—software for mathematical modeling—to find that

$$\frac{df}{dn} = -\frac{2(4n^2 + 5n + 2)}{3(n+1)^2(3n+1)^2} < 0$$

for all $n = 1, 2, 3, \ldots$. A simple *Maple* code and the results of differentiation are shown in Figure 1.16. Therefore, $f(n)$ monotonically decreases for $n \geq 1$ and, thereby, $f(n) \leq f(1) = \frac{5}{8}$ and, as simple algebraic transformations can show, $f(n) > \frac{4}{9}$. Because $v(n) = \frac{f(n)}{n+1}$ it follows

$$\frac{4}{9(n+1)} < v(n) \leq \frac{5}{8(n+1)} \tag{27}$$

Finally, as $n + 1 \leq 2n$ and $n + 1 > n$ it follows that $\frac{4}{9(n+1)} \geq \frac{2}{9n}$ and $\frac{5}{8(n+1)} < \frac{5}{8n}$. Therefore, inequalities (27) can be replaced by inequalities (26). This completes the proof.

Proposition 21. The series $\sum_{n=1}^{\infty} v(n)$ diverges.

Proof. It follows from Proposition 17 that $\sum_{n=1}^{\infty} v(n) > \sum_{n=1}^{\infty} \frac{2}{9n} = \frac{2}{9}\sum_{n=1}^{\infty} \frac{1}{n} = +\infty$.

---

[7] In the words of Langtangen and Tveito (2001): "Much of the current focus on algebraically challenging, lengthy, error-prone paper and pencil work can be significantly reduced. The reason for such an evolution is that the computer is simply much better than humans on any theoretically phrased well-defined repetitive operation" (pp. 811-812).

7. Activity Set

1. How many squares with side length measured by an even number are there on the $(2n)\times(2n)$ checkerboard?
2. How many squares with side length measured by an even number are there on the $(2n+1)\times(2n+1)$ checkerboard?
3. How many rectangles with at least one side length measured by an even number can be found on the $(2n)\times(2n)$ checkerboard?
4. How many rectangles with at least one side length measured by an even number can be found on the $(2n+1)\times(2n+1)$ checkerboard?
5. How many rectangles with both sides measured by an even number are there on the $(2n)\times(2n)$ checkerboard?
6. How many rectangles with both sides measured by an even number are there on the $(2n+1)\times(2n+1)$ checkerboard?
7. Prove that the number of squares on the $n \times m$ checkerboard is equal to $\frac{m(m+1)(3n-m+1)}{6}$.
8. Let $r(n)$ be the ratio of the number of squares with side length measured by an odd number to the total number of rectangles on the $(2n)\times(2n)$ checkerboard.

   Find $\lim_{n\to\infty} r(n)$. Evaluate $\sum_{n=1}^{\infty} r(n)$ (that is, decide whether the series converges or diverges).
9. Let $r(n)$ be the ratio of the number of rectangles with at least one side measured by an even number to the total number of rectangles on the $n\times n$ checkerboard.

   Find $\lim_{n\to\infty} r(n)$. Evaluate $\sum_{n=1}^{\infty} r(n)$.
10. Let $r(n)$ be the ratio of the number of rectangles with both sides measured by an even number to the total number of rectangles on the $n\times n$ checkerboard.

    Find $\lim_{n\to\infty} r(n)$. Evaluate $\sum_{n=1}^{\infty} r(n)$.
11. How many rectangles with side lengths in the ratio two to one are there on the $2n\times 2n$ checkerboard?
12. How many rectangles with side lengths in the ratio two to one are there on the $(2n+1)\times(2n+1)$ checkerboard?
13. Let $r(n)$ be the ratio of the number of rectangles with side lengths in the ratio two to one to the total number of rectangles on the $(2n)\times(2n)$ checkerboard.

    Find $\lim_{n\to\infty} r(n)$. Evaluate $\sum_{n=1}^{\infty} r(n)$.
14. Let $r(n)$ be the ratio of the number of rectangles with side lengths in the ratio two to one to the total number of rectangles on the $(2n+1)\times(2n+1)$ checkerboard.

    Find $\lim_{n\to\infty} r(n)$. Evaluate $\sum_{n=1}^{\infty} r(n)$.

Chapter 2

ALGEBRAIC EQUATIONS WITH PARAMETERS

Plato was asked, "What does God do?" and had to reply, "God eternally geometrizes."
— Reid (1963, p. 1).

1. INTRODUCTION

One of the core recommendations of the Conference Board of the Mathematical Sciences (2001) for the preparation of teachers focuses on the need for courses based on the exploration of fundamental mathematical concepts that leads to the development of the mindset of a mathematician. In addition, these courses should take full advantage of increasingly sophisticated technological tools that enable such explorations. Solving equations in one variable is a traditional topic in secondary school algebra. Typically, learners of mathematics do not have difficulties with this topic. For example, in the case of quadratic equations the pre-college curriculum centers on the development of skills in either factoring a trinomial or using the quadratic formula. Unfortunately, such treatment of the topic pays little attention to one's conceptual development. The reformed vision of secondary school algebra goes beyond the need for memorizing formula and mastering factorization techniques. Nowadays, algebra can and must be taught through an experimental approach as a dynamic and conceptually oriented subject matter, permeated by the exploration of computer-generated geometric representations of algebraic models that leads to conjecturing and, ultimately, to formal demonstration of mathematical propositions so discovered. In the case of equations with parameters, the appropriate use of technology can provide conceptually oriented learning environments conducive to the development of advanced mathematical thinking.

This chapter will demonstrate that significant curricular implications may result from shifting the focus of school mathematics activities from the study of specific equations to those dependent on parameters. The introduction of parameter-dependent functions and equations in the curriculum has great potential to bring about a dynamic flavor in the seemingly static structure of pre-calculus and, in general, supports the reformed vision of school mathematics. Indeed, when exploring equations with parameters, one can shift focus from the search for numbers that solve a particular equation to the study of the structural properties of a family of equations that provide the solutions of a specified type. This kind of mathematical behavior, resembling professional activities of mathematicians and, more generally, STEM workforce, has great potential to reorganize mathematics classrooms according to the vision expressed by the National Council of Teachers of Mathematics (2000): "Imagine a classroom … [where] technology is an essential component of the environment [and] students confidently engage in complex mathematical tasks chosen carefully by teachers" (p. 3). In other words, such a pedagogical position calls for both the change of curricula and re-conceptualization of traditional teaching strategies. These changes, in turn, require new topics to be included in mathematics education courses for teachers. In what follows, a number of pedagogical ideas that have the potential to enhance an exploratory, computer-enhanced introduction of traditional and advanced topics in algebra will be provided. These ideas reflect the author's work with teachers in a capstone course using the Graphing Calculator 3.5 (*GC*).

2. A Locus Approach to Quadratic Equations with Parameters

A locus is a set of points determined by a specified condition applied to a function. Consider the quadratic function $f(x,c) = x^2 + x + c$ of variable x with parameter c. One can say that the graph of the equation

$$x^2 + x + c = 0 \tag{1}$$

is a locus defined by the zero value for the function $f(x,c)$. How does the graph of equation (1) in the plane (x, c) look like? To answer this question, one can rewrite equation (1) in the form $c = -(x^2 + x)$ and then, using conventional notation, construct the graph $y = -(x^2 + x)$. The use of the *GC* makes it possible to construct the graph of equation (1) directly without representing parameter c as the function of x. Regardless of the type of graphing software used, the locus of equation (1), as shown in Figure 1, is a parabola open downwards with x-intercepts at the points $x = -1$ and $x = 0$. This parabola can be used as a tool for answering many questions about the roots of quadratic equation (1) *without* having a formula that solves this equation. In fact, the very formula can be derived from the inquiry into the properties of the locus. Following are examples of eight explorations that can be carried out in the context of equation (1) using the locus approach enhanced by the *GC*.

Exploration 1. For what values of parameter c does equation (1) have two real roots?

Reflections. A traditional approach to answering this question involves the use of the quadratic formula (see formula (5) below) followed by setting the discriminant inequality $1 - 4c \geq 0$ which yields $c \leq 0.25$. However, the last inequality can be directly derived from Figure 1 if one interprets the roots of equation (1) as the x-intercepts of the locus and the horizontal line c = *constant*. Indeed, the two lines intersect only when this *constant* is not greater than 0.25. That is, for all $c < 0.25$ equation (1) has two real roots and when $c = 0.25$ equation (1) has a double root, $x = -0.5$.

Exploration 2. For what values of parameter c does equation (1) have two positive roots?

Reflections. Note that in order to answer this question through the locus approach, one does not need to construct a series of graphs $y = x^2 + x + c$ for different values of parameter c (using graphing technology) or to carry out transformation of inequalities involving radicals that result from the use of the quadratic formula. Quite the contrary, by analyzing the locus of equation (1) pictured in Figure 1, one can conclude that because no horizontal line c = *constant* can intersect the locus to the right of the origin only, such values of c do not exist. As an aside, note that hereafter, when typing an equation with parameter in the equation window, the variable y will be used in place of the parameter (in this case, y is substituted for c).

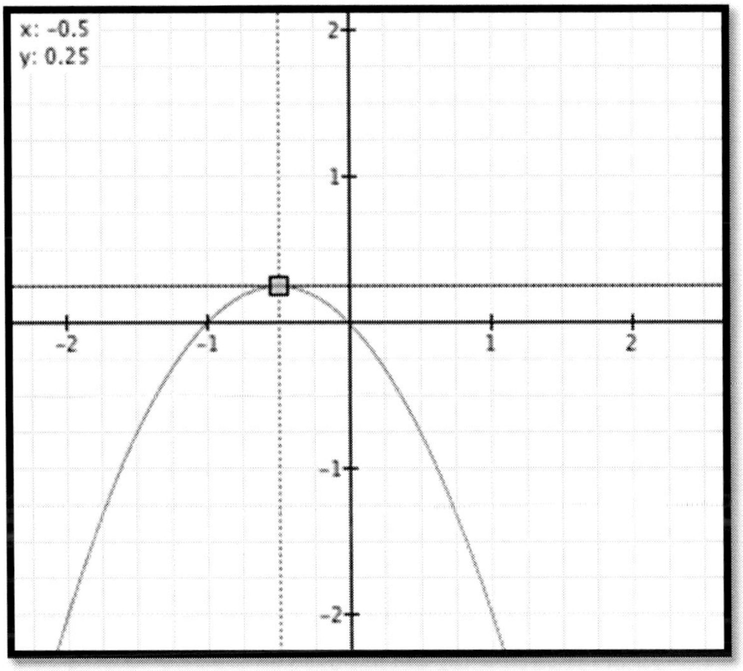

Figure 2.1. Locus of equation (1).

Exploration 3. For what values of parameter c does equation (1) have two negative roots?

Reflections. The property of equation (1) of having two negative roots is equivalent to the intersection of the line y = *constant* with the locus in two points with negative x-coordinates. It follows from Figure 1 that any $c \in (0, 0.25)$ allows for such an intersection thus providing equation (1) with two negative roots. This completes the exploration.

Exploration 4. For what values of parameter c does equation (1) have roots located by different sides of the origin?

Reflections. Once again, the locus approach enables one to complete the exploration in a single step: for all $c < 0$ the locus and a horizontal straight line (c = -0.6 in Figure 1) intersect twice and at the different sides of the origin, implying the existence of two real roots of different signs.

Exploration 5. Find the sum of the roots of equation (1).

Reflections. One can see (Figure 1) that the line x = -0.5 is the line of symmetry of the locus of equation (1). Therefore, the x-intercepts of the points in common of the locus and the line c = *constant* (alternatively, the roots of equation [1]) are equidistant from the point x = -0.5 and, thereby, can be expressed in the form

$$x_1 = -0.5 - R, \quad x_2 = -0.5 + R \qquad (2)$$

Here R is a positive number that characterizes equal distance from the points (x_1, c) and (x_2, c) of the locus to the line of symmetry. It follows from relations (2) that

$$x_1 + x_2 = -1 \quad (3)$$

One may note that the sum of the roots of equation (1) is the negation of the coefficient in x. Alternatively, given $c = c^*$, one may note that $x = -0.5$ is the x-coordinate of the midpoint of the segment connecting the points (x_1, c^*) and (x_2, c^*) of the locus. Therefore,

$$\frac{x_1 + x_2}{2} = -0.5 .$$

Multiplying both sides of the last relation by two yields (3).

Exploration 6. Find the product of the roots of equation (1).

Reflections. From relation (3) it follows that $x_2 = -1 - x_1$. Furthermore, $x_1^2 + x_1 + c = 0$, as x_1 is a root of equation (1). Therefore,

$$x_1 x_2 = x_1(-1 - x_1) = -(x_1^2 + x_1) = -(x_1^2 + x_1 + c - c) = -(0 - c) = c .$$

That is,

$$x_1 x_2 = c \quad (4)$$

Remark 1. Relations (3) and (4), expressing the sum and the product of the roots of an equation through its coefficients, are known as Viète's[1] formulas for quadratic equation (1).

Exploration 7. Develop a formula expressing the roots of equation (1) through the coefficients of this equation.

Reflections. Using relations (2) and (4) one can write

$$c = (-0.5 - R)(-0.5 + R) = (-0.5)^2 - R^2 = 0.25 - R^2$$

whence $R = \sqrt{\frac{1}{4} - c}$. Therefore, $x_1 = -\frac{1}{2} - \sqrt{\frac{1}{4} - c}$ and $x_2 = -\frac{1}{2} - \sqrt{\frac{1}{4} - c}$. Simplifying the last two relations yields the formula

$$x_{1,2} = \frac{-1 \pm \sqrt{1 - 4c}}{2} \quad (5)$$

[1] François Viète – a French mathematician of the 16th century who was the first to introduce the use of letters in algebra.

Formula (5), expressing x_1 and x_2 through the coefficients of equation (1), is known as the quadratic formula for equation (1).

Remark 2. Using the *GC* one can graph in the plane (x, c) the relations

$$x = \frac{-1-\sqrt{1-4c}}{2} \text{ and } x = \frac{-1+\sqrt{1-4c}}{2}$$

to get, respectively, the left and right branches of the locus (parabola) of equation (1) shown in Figure 1.

Remark 3. Whereas the formula oriented approach to quadratic equations is based on memorization and the use of algebraic symbolism, the locus approach treats the locus of the family of quadratic equations as an object to be studied and a tool to think with. This orientation emphasizes the conceptual understanding of algebraic structures rather than the development of skills in the use of algebraic transformations alone. It leads naturally to more advance learning situations such as the study of the roots' location about an arbitrary point (rather than about the origin only).

Exploration 8. For what values of parameter c does equation (1) have two different roots separated by the number 1?

Reflections. A traditional approach to this task using computer graphing software involves the construction of the series of graphs of the function $y = x^2 + x + c$ until the value of c that provides the family of the graphs with an x-intercept equal to the number 1 is located. Then one can see that as c becomes smaller than -2, the roots come to be separated by 1.The approach based on the graphing of a series of parabolas can be significantly improved through the locus approach as the latter involves the construction of the locus of equation (1) and the vertical line $x = 1$ enabling one to see that the two graphs intersect at $c = -2$ (Figure 2.2). Furthermore, one can see in a single graph that for all $c < -2$ the corresponding points of the locus are located by different sides of the vertical line $x = 1$. This implies that for $c < -2$ the number 1 resides between the corresponding roots (the x-intercepts of points in common of the line $c = constant$ and the locus).

Remark 4. A pure analytical approach to Exploration 8 would have required using formula (5), choosing the largest root, and then solving the inequality $\frac{-1+\sqrt{1-4c}}{2} > 1$ which is equivalent to the inequality $\sqrt{1-4c} > 3$ whence $c < -2$. Note that the last transition is not obvious. Indeed, in the case of the inequality with the opposite sign, $\sqrt{1-4c} < 3$, the squaring of the both sides without taking into account the domain of the inequality would lead to an incorrect solution, $c > -2$, rather than $-2 < c < 0.25$. However, the locus approach enables one to avoid dealing with the peculiarities of an analytical solution of an irrational inequality.

The locus approach makes it possible to examine the structure of a quadratic equation by using essentially a single graph (Figure 2). Moreover, the locus approach enables one to extend Exploration 8 to a multitude of quadratic equations. For example, plugging $x = 2$ into the equation $2x^2 + x + c = 0$ yields $8 + 2 + c = 0$ whence $c = -10$. Thus, one can conclude that for all $c < -10$ the last quadratic equation has two real roots separated by the number 2 .

Apparently, the change of coefficient in x^2 (or in x) in equation (1) yields a new problem, yet such a change does not require the construction of a new locus. One should be encouraged to work on several tasks involving same type of locus, in order to develop and use its mental image as a thinking device. Just as the knowledge of the commutative property of multiplication allows one to avoid the use of a calculator in multiplying two numbers if they were already multiplied in different order, the mental image of a locus enables one to practice graphing skills in the absence of technology. Therefore, the locus approach has a potential to create a residual mental power that can be used by learners in a technology-free setting. Such an approach to using technology is pedagogically-appropriate in the sense that it includes the joint use of computer-enhanced explorations and follow-up tasks that are technology-free by design.

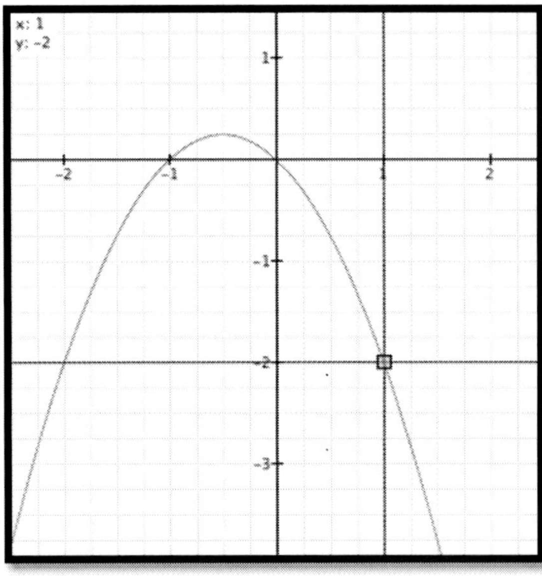

Figure 2.2. Exploration 8 and the locus approach.

3. Quadratic Equations with a Hyperbola-Like Locus

The explorations carried out so far involved only one family of quadratic equations, and, whereas the locus of equation (1)can provide a sufficiently rich problem-solving milieu, the locus approach can be naturally extended to include other families of quadratic equations. To this end, consider more complicated phenomena exhibited by the family of equations of the form

$$x^2 + bx - 1 = 0 \quad (6)$$

where b is a real parameter. The locus of equation (6) consists of two hyperbolas with the asymptotes $x = 0$ and $y = - x$ shown in Figure 2.3. In what follows, seven explorations related to equation (6) will be discussed.

Exploration 9. For what values of parameter b does equation (6) have real roots?

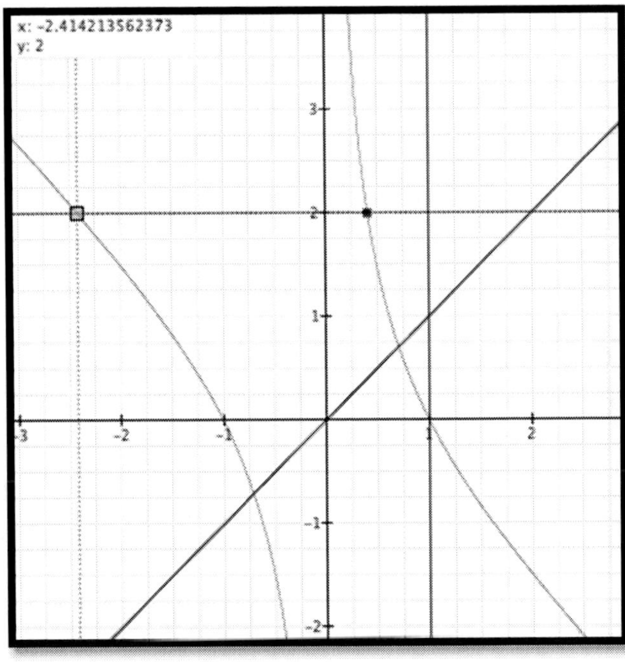

Figure 2.3. A hyperbola-like locus of equation (6).

Reflections. The locus of equation (6) when pictured jointly with the horizontal line $b =$ *constant* (Figure 2.3) enables one to see that the two graphs always have two points in common. Analytically, this geometric phenomenon can be interpreted as follows: equation (6) has real roots for all real values of parameter b.

Exploration 10. For what values of parameter b does equation (6) have roots of different signs?

Reflections. Once again, one can observe (Figure 2.3) that the locus of equation (6) consists of two branches located by the different sides of the (vertical) b-axis. Analytically, this observation means that the line $b =$ *constant* has a positive x-intercept with the right branch and a negative x-intercept with the left branch. One can conclude that for any value of b equation (6) has roots of different signs.

Exploration 11. For what values of parameter b does equation (6) have both roots smaller (greater) than the number 1?

Reflections. The locus of equation (6) and the line $x = 1$ (Figure 2.3) enable one to see that for all $b > 0$ the line $b =$ *constant* always intersects the locus at two points with x-coordinates smaller than one. This kind of behavior of graphs suggests that for any positive value of b equation (6) has both roots smaller than 1 and no value of b exists for which both roots are greater than 1.

Remark 5. Once again, the locus approach makes it possible to decide the location of the roots of a quadratic equation about a given point without knowing how to calculate the values of the roots. In order to really appreciate the power of the locus approach, one can try to answer questions posed in Exploration 11 using the quadratic formula for equation (6) and techniques of solving irrational inequalities.

Exploration 12. Let x_1 and x_2 be real roots of equation (6). For what values of parameter b do the inequalities $x_1 < x_2 < b$ (alternatively, $b < x_1 < x_2$ or $x_1 < b < x_2$) hold true?

Reflections. Consider the first pair of inequalities. One has to compare the values of x located on the x-axis to the value of b located on the b-axis. How can the coordinates of points that belong to different axes be compared? To this end, one has to use a tool that allows one to map any point from the b-axis to the x-axis and vice versa. Such a tool is the line $b = x$—the bisector of the first and the third coordinate angles. Figure 2.4 shows three graphs: the locus of equation (6), the horizontal line b = *constant*, and the bisector $b = x$. Therefore, the x-intercepts of the line b = *constant* with two branches of the locus and the bisector $b = x$ have to be compared.

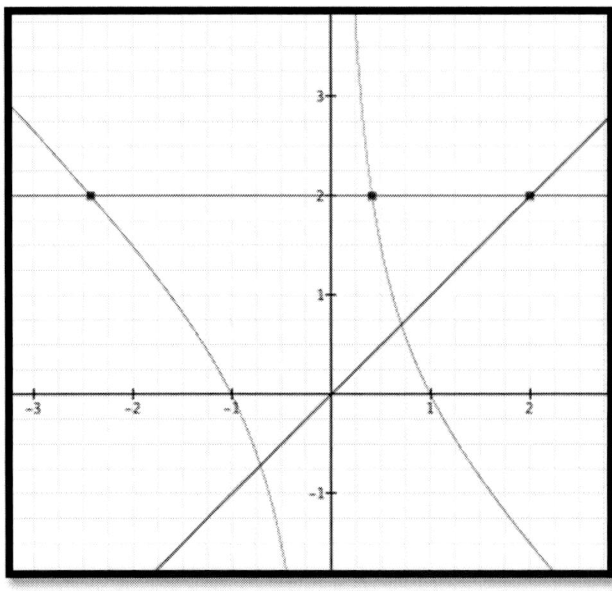

Figure 2.4. Mapping parameter b to the x-axis.

b	x_1	x_2	$x_1 + x_2$
1	-1.61083...	0.61083...	-1
2	-2.41421...	0.241421...	-2
3	-3.30277...	0.30277...	-3
4	-4.23606...	0.23606...	-4

Figure 2.5. A computational approach to Exploration 13.

The inequalities $x_1 < x_2 < b$ imply that the line $b = x$ is the last one to be crossed by the line b = *constant.* As follows from Figure 2.4, for all values of $b > b^*$, where b^* is the positive root of equation (6) when $x = b$, the inequalities $x_1 < x_2 < b$ hold. Substituting x for b

in equation (6) yields $2b^2 - 1 = 0$ whence $b^* = \frac{1}{\sqrt{2}}$. That is, for all $b > \frac{1}{\sqrt{2}}$ the inequalities $x_1 < x_2 < b$ hold. A similar exploration of the other two pairs of inequalities yields that for $b < -\frac{1}{\sqrt{2}}$ the inequalities $b < x_1 < x_2$ hold and for $-\frac{1}{\sqrt{2}} < b < \frac{1}{\sqrt{2}}$ the inequalities $x_1 < b < x_2$ hold.

Exploration 13. Find the sum of the roots of equation (6).

Reflections. Consider a few pairs of the x-intercepts of the line b = *constant* with the locus of equation (6). As shown in Figure 2.3, for $b = 2$ we have $x_1 = -2.41421\ldots$ and $x_2 = 0.41421\ldots$ whence $x_1 + x_2 = -2$. In much the same way other pairs of roots can be collected in the table shown in Figure 2.5.

Comparing the values in the far-right and far-left columns of the table, the following computationally motivated formula results

$$x_1 + x_2 = -b \qquad (7)$$

In other words, just like in the case of equation (1), the sum of the roots of equation (6) is equal to the negation of the coefficient in x (often called the second coefficient[2]).

Exploration 14. Find the product of the roots of equation (6).

Reflections. The data presented in the table of Figure 2.5 allows one to find the product of the roots computationally in all four cases of parameter b:

$$x_1x_2\,|_{b=1} = x_1x_2\,|_{b=2} = x_1x_2\,|_{b=3} = x_1x_2\,|_{b=4} = -1$$

This result can be confirmed analytically by using formula (7). Indeed, substituting $x_1 = -b - x_2$ for x in the left-hand side of equation (6) yields

$$(-b-x_2)^2 + b(-b-x_2) - 1 = b^2 + 2bx_2 + x_2^2 - b^2 - bx_2 - 1 = x_2^2 + bx_2 - 1 = 0.$$

Therefore,

$$x_1x_2 = -1 \qquad (8)$$

In other words, the product of the roots of equation (6) is equal to its free term[3].

---

[2] It is interesting to note that in the general case of the equation of the n-th degree $x^n + bx^{n-1} + cx^{n-2} + \ldots = 0$, we also have $x_1 + x_2 + \ldots + x_n = -b$ (thus the term "the second coefficient"). For example, the equation $x^5 + 10x^4 + 36x^3 + 56x^2 + 35x + 6 = 0$ has five real roots the sum of which is equal to -10. One can verify this fact by graphing the equation with the *GC* and finding the x-intercepts (of the resulting five vertical lines) through cursor pointing

[3] Using the *GC*, one can check to see that the product of the (five real) roots of the equation $x^5 + 10x^4 + 36x^3 + 56x^2 + 35x + 6 = 0$ is equal to -6. In the general case of the equation $x^n + bx^{n-1} + cx^{n-2} + \ldots + r = 0$ we have $x_1 \cdot x_2 \cdot \ldots \cdot x_n = (-1)^n r$.

Remark 6. Relations (7) and (8), expressing the sum and the products of the roots of an equation through its coefficients, are known as Viète's formulas for quadratic equation (6).

Exploration 15. Use the graph of the line $b=-2x$ jointly with the locus of equation (6) in order to develop a formula for the roots x_1 and x_2 of this equation.

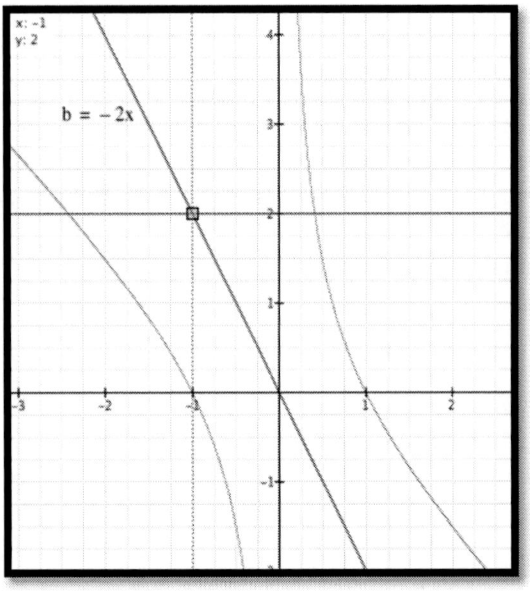

Figure 2.6. Quasi-symmetry of the locus of equation (6).

Reflections. As shown in Figure 2.6, the locus of equation (6) appears to be symmetrical about the line $b=-2x$. The point in common of the graphs $b=-2x$ and b = *constant* (in Figure 2.6, b = 2), $(-\frac{b}{2},b)$, is the midpoint of the segment connecting the points of intersection (x_1,b) and (x_2,b) of the latter graph with the locus. Therefore, one can set $x_1=-\frac{b}{2}-R$ and $x_2=-\frac{b}{2}+R$, where R is a positive number.

The last two relations and relation (8) imply

$$1=(-\frac{b}{2}-R)(-\frac{b}{2}+R)=R^2-\frac{b^2}{4}$$

whence $R=\sqrt{\frac{b^2}{4}+1}$ and, thereby, $x_1=-\frac{b}{2}-\sqrt{\frac{b^2}{4}+1}$ and $x_2=-\frac{b}{2}+\sqrt{\frac{b^2}{4}+1}$. Finally, the formula for the roots of equation (6) can be simplified to the form

$$x_{1,2}=\frac{-b\pm\sqrt{b^2+4}}{2} \qquad (9)$$

Remark 7. Using the *GC*, one can graph the equations $x = \frac{-b - \sqrt{b^2 + 4}}{2}$ and $x = \frac{-b + \sqrt{b^2 + 4}}{2}$ in the plane (x, b) to get, respectively, the left and right branches of the locus as shown in Figures 2.3, 2.4, and 2.6.

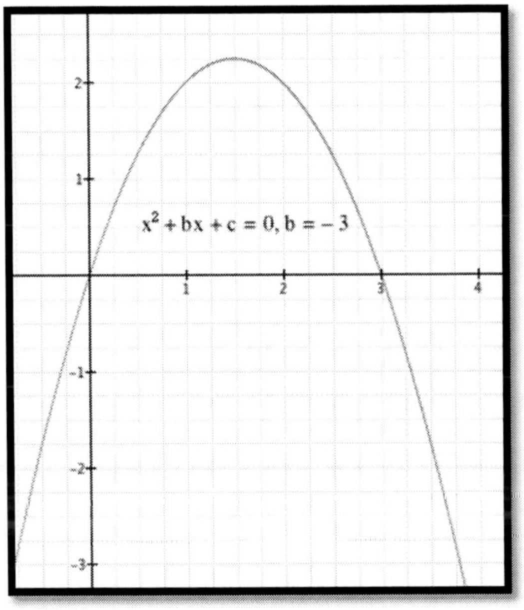

Figure 2.7. A loci of equation (10) in the plane (x, c) for $b = 3$.

4. Quadratic Equations with Two Parameters

Consider the quadratic equation

$$x^2 + bx + c = 0 \qquad (10)$$

Note that the case of equation (10) is the most general one as the equation $Ax^2 + Bx + C = 0$, $A \neq 0$, with three parameters can be reduced to equation (10) with $b = \frac{B}{A}$ and $c = \frac{C}{A}$. How can one use the locus approach in the case of two parameters? One can first construct a series of loci of equation (10) in the plane (x, c) for different values of b (controlling the variation of b by a slider). Regardless of the value of b, the locus is always a parabola open downwards and passing through the origin (Figure 2.7).

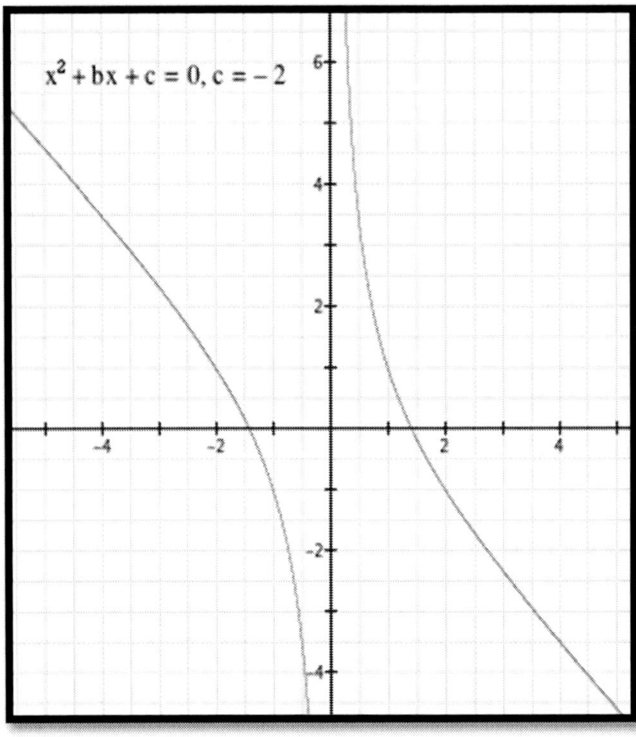

Figure 2.8. The locus of equation $x^2 + bx + c = 0$ in the plane (x, b); $c = -2$.

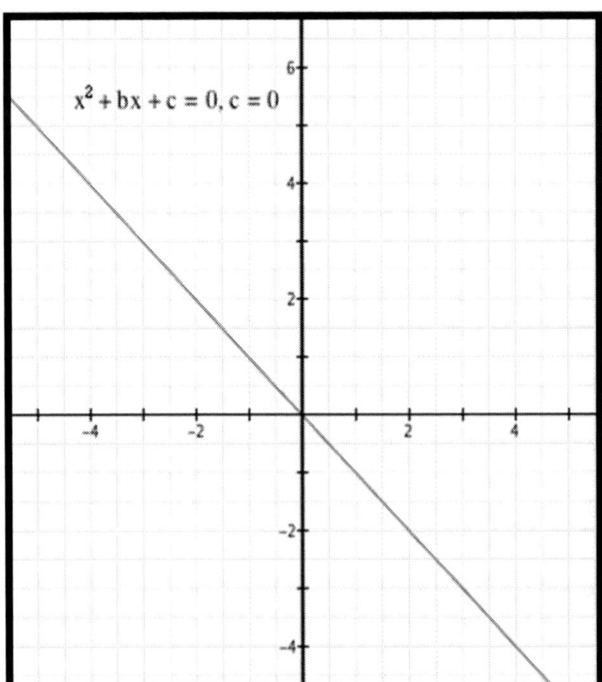

Figure 2.9. The locus of equation $x^2 + bx + c = 0$ in the plane (x, b); $c = 0$.

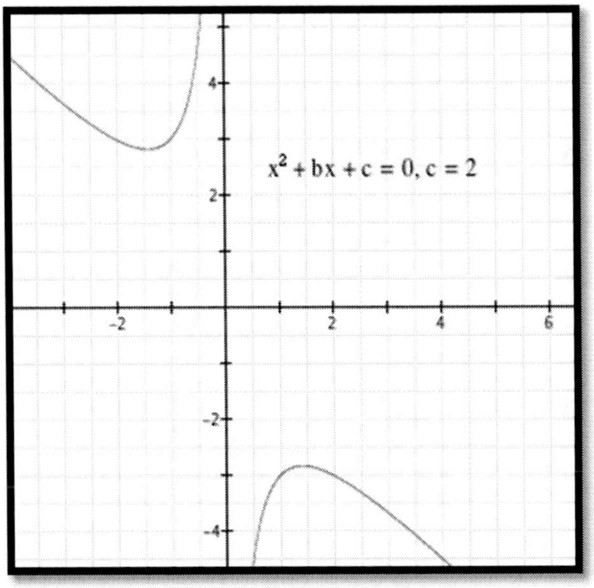

Figure 2.10. The locus of equation $x^2 + bx + c = 0$ in the plane (x, b); $c = 2$.

Then, one can construct a series of loci of equation (10) in the plane (x, b) for different values of c (controlling the variation of c by a slider as well). One can see (Figures 2.8-2.10) that depending on whether $c < 0$, $c = 0$, or $c > 0$, the loci of equation (10) constructed in the plane (x, b) are, respectively, a pair of hyperbolic branches that span through the whole plane (Figure 2.8), a pair of straight lines, $x = 0$ and $b = -x$, (Figure 2.9), or a pair of parabola-like branches where $|b| \geq 2\sqrt{c}$ (Figure 2.10, $|b| \geq 2\sqrt{2}$).

A number of questions can be explored in the context of equation (10) using the loci shown in Figures 2.8-2.10. Those questions are included in the activity set for this chapter. Below, a different kind of locus will be introduced. Rather than constructing a locus in the plane variable-parameter (e.g., $[x, b]$), loci in the plane (b, c) that provide a certain behavior of the graph of the left hand side of equation (10) will be constructed. As the fist example, consider

Exploration 16. Let x_1 and x_2 be real roots of equation (10). For what values of parameters b and c do the inequalities

$$x_1 < c < x_2 \quad (11)$$

$$c < x_1 < x_2 \quad (12)$$

$$x_1 < x_2 < c \quad (13)$$

hold true? In the plane (b, c) construct the loci of inequalities (11)-(13).

Reflections. Consider the function $f(x) = x^2 + bx + c$. Its graph is a parabola open upwards which, depending on b and c, may or may not have points in common with the x-axis. Let us assume that there exist x_1 and x_2, $x_1 < x_2$, such that $f(x_1) = f(x_2) = 0$. Figure 2.11 shows the case of $f(x) = x^2 - 5x + 3$ for which inequalities (11) hold true. One can see, that in general, the inequality $f(c) < 0$ is necessary and sufficient for the roots of equation (10) to satisfy inequalities (11). This yields the inequality $c^2 + bc + c < 0$ the locus of which in the plane (b, c) is shown in Figure 2.12. The locus represents the graph of the two-variable inequality $\sqrt{-(c^2 + bc + c)} > 0$ constructed by the *GC*.

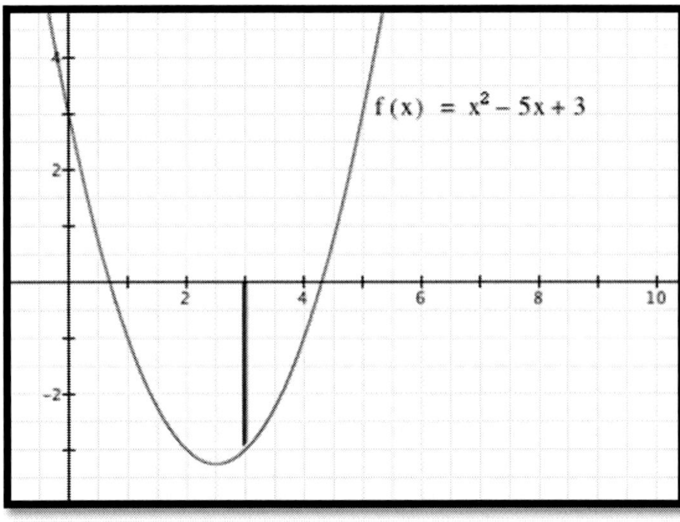

Figure 2.11. The graph of $f(x)$ related to inequalities (11).

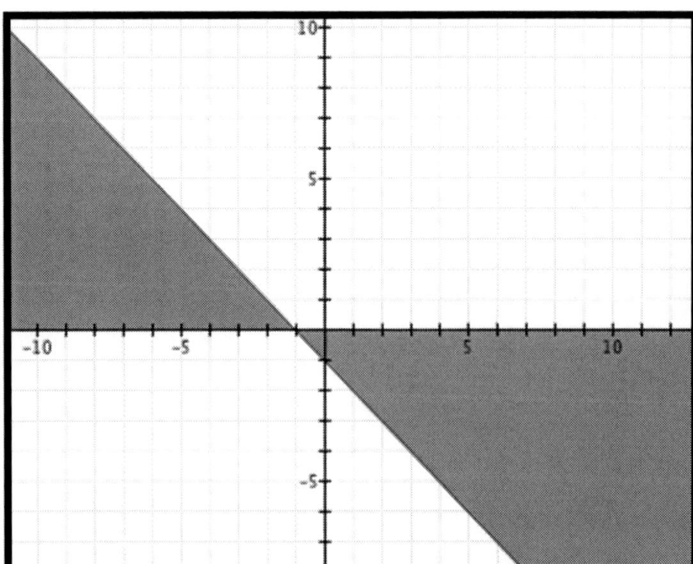

Figure 2.12. The locus of inequalities (11) in the plane (b, c).

Next, Figure 2.13 shows the case $f(x) = x^2 - 2x - 1$ for which inequalities (12) hold true. One can see, that, in general, regardless of the sign of parameter c, the inequalities

$$f(c) > 0\,,\ f(-\frac{b}{2}) < 0\,,\ c < -\frac{b}{2}$$

are necessary and sufficient for the roots of equation (10) to satisfy inequalities (12), where $x = -\frac{b}{2}$ is the line (located to the right of the line $x = c$) of symmetry of the graph of $f(x)$. The last three inequalities can be written as follows

$$c^2 + bc + c > 0,\ b^2 - 4c > 0\,,\ 2c + b < 0 \qquad (14)$$

The locus of the system of simultaneous inequalities (14) in the plane (b, c) is shown in Figure 2.14. Note that the inequality $f(-\frac{b}{2}) < 0$ is an equivalent form of the discriminant inequality—the necessary and sufficient condition of the existence of two real roots of equation (10). As shown in Figure 2.14, the point (-2, 1) belongs to the shaded part of the plane (b, c) and turns equation (10) into the equation $x^2 - 2x - 1 = 0$ whence $x_{1,2} = 1 \pm \sqrt{2}$ yielding the inequalities $-1 < 1 - \sqrt{2} < 1 + \sqrt{2}$ of type (12).

The case of inequalities (13) can be treated in much the same way as the case of inequalities (12) with the only difference in the relationship between the line of symmetry of $f(x)$ and the line $x = c$. That is, the inequalities

$$c^2 + bc + c > 0,\ b^2 - 4c > 0\,,\ 2c + b > 0 \qquad (15)$$

the locus of which is shown in Figure 2.15, are necessary and sufficient for the roots of equation (10) to satisfy inequalities (13). As shown in Figure 2.15, the point (-3, 2.1) belongs to the shaded part of the plane (b, c) and turns equation (10) into the equation $x^2 - 3x + 2.1 = 0$ whence $x_{1,2} = \frac{3 \pm \sqrt{0.6}}{2}$ yielding the inequalities $\frac{3 - \sqrt{0.6}}{2} < \frac{3 + \sqrt{0.6}}{2} < 2.1$ of type (13).

Note that the loci of inequalities (14) and (15) can be constructed using the *GC* as the "graphs" of the two-variable inequalities

$$\sqrt{y^2 + xy + y} \cdot \sqrt{x^2 - 4y} \cdot \sqrt{-(2y + x)} > 0$$

and

$$\sqrt{y^2 + xy + y} \cdot \sqrt{x^2 - 4y} \cdot \sqrt{2y + x} > 0\,,$$

respectively, where $x = b$ and $y = c$. This rather sophisticated construction shows how, in accord with a recommendation of the Conference Board of the Mathematical Sciences (2001) concerning the integration of technology in mathematics education courses for teachers, the appropriate use of available graphing software can demonstrate the power of mathematical definitions.

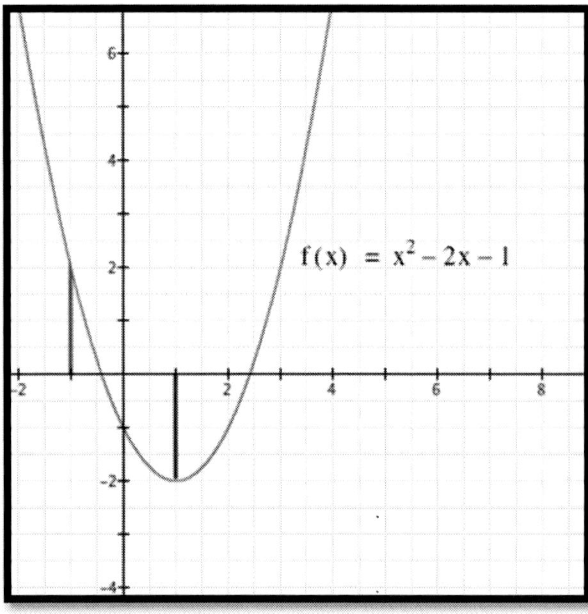

Figure 2.13. The graph of $f(x)$ related to inequalities (12).

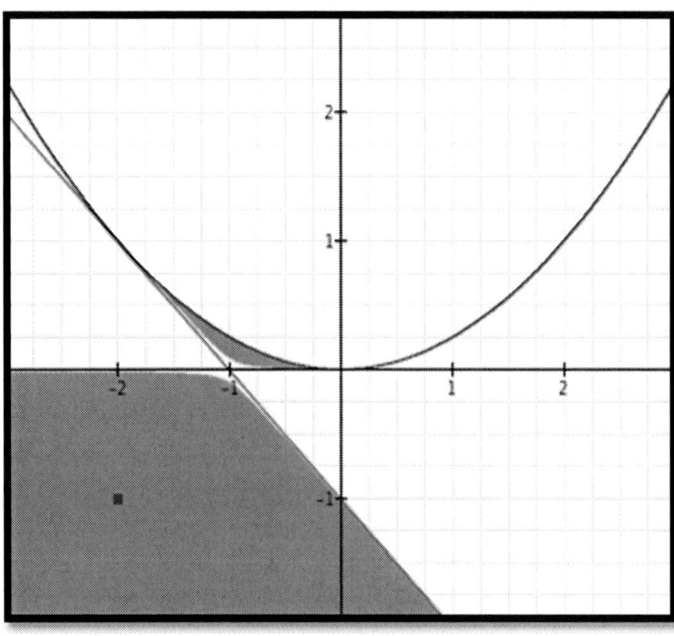

Figure 2.14. The locus of inequalities (12) in the plane (b, c).

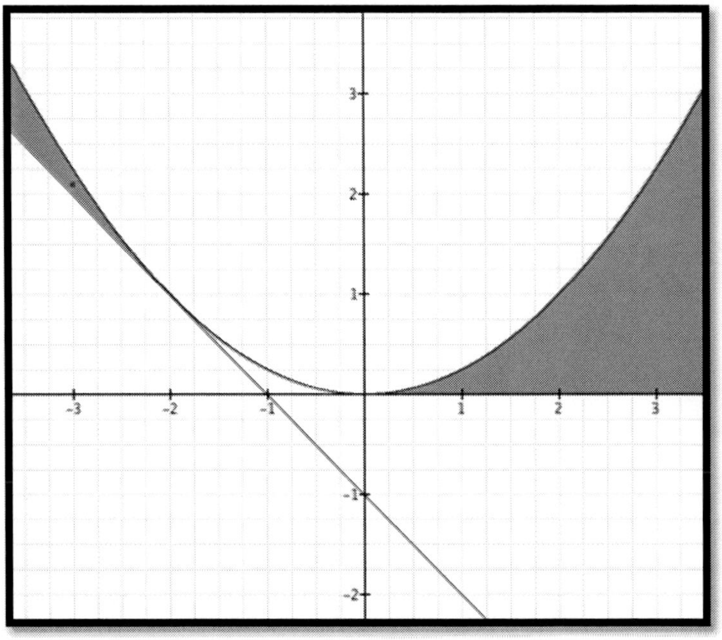

Figure 2.15. The locus of inequalities (13) in the plane (b, c).

Exploration 17. Describe the loci of inequalities (11)-(13) constructed in Figures 2.12, 2.14, and 2.15 analytically in terms of the relationships between the parameters b and c.

Reflections. 1) Consider the locus of inequalities (11) constructed in the plane of parameters (b,c) as the solution to the inequality $c^2 + bc + c < 0$ (the shaded region in Figure 2.12). When c > 0, the coordinate b is not bounded from below and bounded from above by the line $c + b + 1 = 0$ where b assumes the value $-(c + 1)$. In the case $c < 0$, the coordinate b is not bounded from above and bounded from below by the line $c + b + 1 = 0$ where, once again, $b = -(c + 1)$. In terms of inequalities, these observations can be described as follows:

$$-\infty < b < -(c+1), \quad 0 < c < +\infty$$

or

$$-(c+1) < b < +\infty, \quad -\infty < c < 0$$

2) Consider the locus of inequalities (12) constructed in the plane of parameters (b, c) as the solution to the system of simultaneous inequalities (14) (the shaded regions in Figure 2.14). When $c > 0$, the coordinate b is bounded by the lines $c + b + 1 = 0$ and $b^2 - 4c = 0$ where it assumes the values $-(c + 1)$ and $-2\sqrt{c}$, respectively. In the case $c < 0$, the coordinate b is not bounded from below and bounded from above by the line $c + b + 1 = 0$ where, once again, $b = -(c + 1)$. In terms of inequalities, these observations can be described as follows:

$$-(c+1) < b < -2\sqrt{c}, \quad 0 < c < +\infty$$

or

$$-\infty < b < -(c+1), \quad -\infty < c < 0 \tag{16}$$

In particular, one can see that the point (-2, -1) is a solution to the second pair of inequalities (16) and, as was mentioned above, the graph of the function y = $x^2 - 2x - 1$ has two x-intercepts satisfying inequalities (12).

3) Finally, consider the locus of inequalities (13) constructed in the plane of parameters (b, c) as the solution to the simultaneous inequalities (15) (the shaded regions in Figure 2.15). When $c > 1$, the coordinate b belongs to two regions: in the case $b < 0$ it is bounded from below by the line $c + b + 1 = 0$ and from above by the line $b^2 - 4c = 0$ where b assumes the values $-(c + 1)$ and $-2\sqrt{c}$, respectively; when $b > 0$ this coordinate is not bounded from above and is bounded from below by the line $b^2 - 4c = 0$ where $b = 2\sqrt{c}$. In the case $0 < c < 1$, the coordinate b belongs to one region only where, once again, is not bounded from above and is bounded from below the line $b^2 - 4c = 0$ where $b = 2\sqrt{c}$. When c < 0, no points belong to the locus of inequalities (13). In terms of inequalities, these observations can be described as follows:

$$-(c+1) < b < -2\sqrt{c}, c > 1; \; 2\sqrt{c} < b < +\infty, c > 1;$$

or

$$2\sqrt{c} < b < +\infty, 0 < c < 1 \tag{17}$$

In particular, the point (-3, 2.1) is a solution to the first pair of inequalities (17) and, as was mentioned above, the graph of function $y = x^2 - 3x + 2.1$ has two x-intercepts satisfying inequalities (13).

Exploration 18. Let x_1 and x_2 be the roots of equation (10). Determine the region in the plane of parameters (c, b) for which the inequalities

$$b < x_1 < x_2 \tag{18}$$

hold true.

Reflections. Similarly to Exploration 16, the inequalities

$$f(b) > 0, \quad f(-\frac{b}{2}) < 0, \quad b < -\frac{b}{2}$$

can be established as necessary and sufficient conditions for inequalities (18) to hold true.

These conditions can be re-written as the system of inequalities in terms of b and c as follows:

$$2b^2 + c > 0, \quad b^2 - 4c > 0, \quad b < 0 \tag{19}$$

Figure 2.16 shows the shaded region (locus) in the plane of parameters (*c*, *b*) where simultaneous inequalities (19) take place. The locus can be constructed using the *GC* as the graph of the two-variable inequality $\sqrt{2y^2+x}\sqrt{y^2-4x}\sqrt{-y}>0$. This region can be described as either

$$-2b^2 < c < \frac{b^2}{4},\ b< 0$$

or

$$-\infty < b < -\sqrt{-\frac{c}{2}},\ c< 0\ ;\ -\infty < b < -2\sqrt{c},\ c> 0.$$

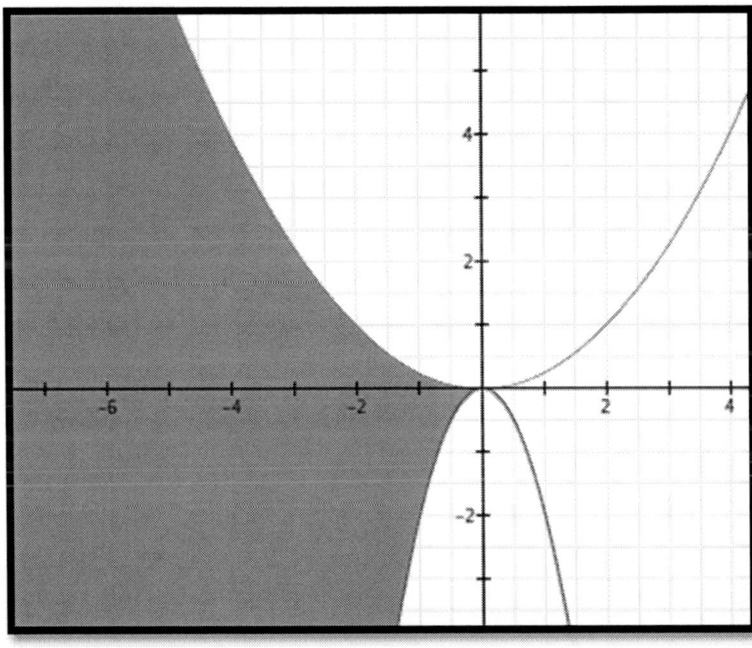

Figure 2.16. The locus of inequalities (19) in the plane (*c*, *b*).

5. Cubic Equations with Two Parameters

In this section, several types of equations of the third degree will be explored using the locus approach. We begin with the equation

$$x^3 + px + q = 0 \tag{20}$$

where p and q are real parameters. First of all note that any cubic equation

$$y^3 + ay^2 + by + c = 0$$

can be reduced to (20) by the substitution $y = x - \frac{a}{3}$. Indeed,

$$y^3 + ay^2 + by + c = (x - \frac{a}{3})^3 + a(x - \frac{a}{3})^2 + b(x - \frac{a}{3}) + c$$
$$= x^3 - x^2 a + xa^2 - \frac{a^3}{27} + ax^2 - \frac{2}{3}xa^2 + \frac{a^3}{9} + bx - \frac{ba}{3} + c$$
$$= x^3 + (\frac{a^2}{3} + b)x + (\frac{2a^3}{27} - \frac{ba}{3} + c) = x^3 + px + q,$$

where $p = \frac{a^2}{3} + b,\ \ q = \frac{2a^3}{27} - \frac{ba}{3} + c$. Therefore, in exploring cubic equations, equation (20) can be considered without loss of generality.

Exploration 19. For various (positive and negative) values of parameter p draw the locus of equation (20) in the plane (x, q). Explore the dependence of the form of the locus on parameter p.

Reflections. Observing Figure 2.17 one can conclude that whereas the locus of equation (20) always intersects the x-axis at the origin regardless of the sign of parameter p, for $p > 0$ the origin is the only such intersection and for $p < 0$ the locus has two more points in common with the x-axis. This implies that for $p > 0$, whatever the value of q, equation (20) has only one real root. At the same time, for $p < 0$, depending on the value of q, equation (20) may have one, two, or three (different) real roots.

Using techniques of calculus, one can explore equation (20) in the case $p < 0$. To this end, consider the function $q(x) = -x^3 - px$ with parameter p. Its derivative, $q'(x) = -3x^2 - p$, has two roots: $x_1 = \sqrt{\frac{-p}{3}}$ and $x_2 = -\sqrt{\frac{-p}{3}}$ so that

$$q_1 = q(x_1) = -(\sqrt{\frac{-p}{3}})^3 - p\left(\sqrt{\frac{-p}{3}}\right) = -\frac{2p}{3}\sqrt{\frac{-p}{3}}$$

and

$$q_2 = q(x_2) = -(-\sqrt{\frac{-p}{3}})^3 - p\left(-\sqrt{\frac{-p}{3}}\right) = \frac{2p}{3}\sqrt{\frac{-p}{3}}.$$

Furthermore, in the neighborhood of x_2, the derivative $q'(x)$ changes its sign from minus to plus and in the neighborhood of x_1—from plus to minus. That is, q_2 and q_1 are, respectively, the local minimum and the local maximum of the function $q(x)$.

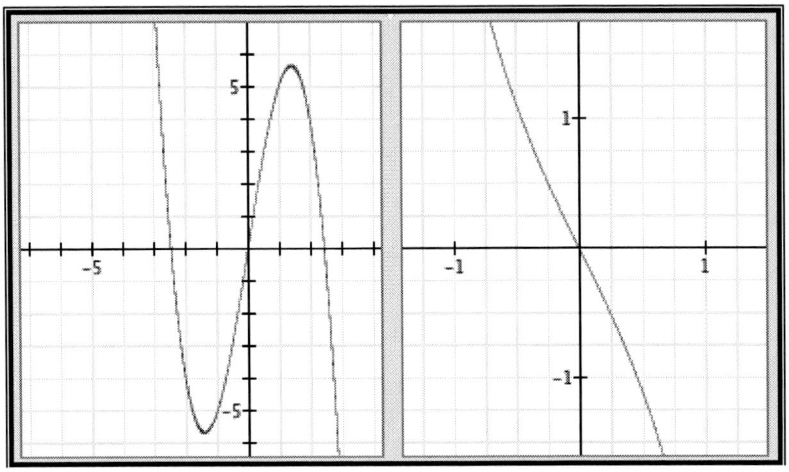

Figure 2.17. Locus of equation (20) in the plane (x, q): left—$p < 0$; right—$p > 0$.

In that way, when $p < 0$ for all $|q| < -\frac{2p}{3}\sqrt{\frac{-p}{3}}$ equation (20) has three real roots, for $|q| = -\frac{2p}{3}\sqrt{\frac{-p}{3}}$ equation (20) has two real roots, and for $|q| > -\frac{2p}{3}\sqrt{\frac{-p}{3}}$ equation (20) has one real root only. One can partition the plane of parameters (p, q) into regions where equation (20) has different number of real roots. For example, the shaded region in Figure 2.18 shows those values of parameters p and q for which equation (20) has three real roots.

In particular, one can check to see that when $(p, q) = (-6, -4)$, the equation $x^3 - 6x - 4 = 0$ has three real roots. Indeed,

$$\begin{aligned} x^3 - 6x - 4 &= x^3 + 8 - 6x - 12 = (x+2)(x^2 - 2x + 4) - 6(x+2) \\ &= (x+2)(x^2 - 2x - 2) = 0 \end{aligned}$$

whence $x_1 = -2, x_{2,3} = 1 \pm \sqrt{3}$.

Note that $x_1 + x_2 + x_3 = -2 + 1 + \sqrt{3} + 1 - \sqrt{3} = 0$—the coefficient in x^2, $x_1x_2 + x_1x_3 + x_2x_3 = (-2)(1+\sqrt{3}) + (-2)(1-\sqrt{3}) + (1+\sqrt{3})(1-\sqrt{3}) = -6$—the coefficient in x, and $x_1x_2x_3 = (-2)(1+\sqrt{3})(1-\sqrt{3}) = 4$—the negation of the free term of the equation $x^3 - 6x - 4 = 0$.

By the same token, when $(p, q) = (-2, 4)$, the equation $x^3 - 2x + 4 = 0$ has one real root only. Indeed,

$$\begin{aligned} x^3 - 2x + 4 &= x^3 + 8 - 2x - 4 = (x+2)(x^2 - 2x + 4) - 2(x+2) \\ &= (x+2)(x^2 - 2x + 2) = 0 \end{aligned},$$

whence $x_1 = -2, x_{2,3} = 1 \pm i$. However, one still has $x_1 + x_2 + x_3 = 0$—the coefficient in x^2, $x_1x_2 + x_1x_3 + x_2x_3 = -2$—the coefficient in x, and $x_1x_2x_3 = -2(1+i)(1-i) = -4$—the negation of the free term of the equation $x^3 - 2x + 4 = 0$. These relations, connecting roots and coefficients of the cubic equation, are special cases of Viète's formulas for a polynomial equation with one variable.

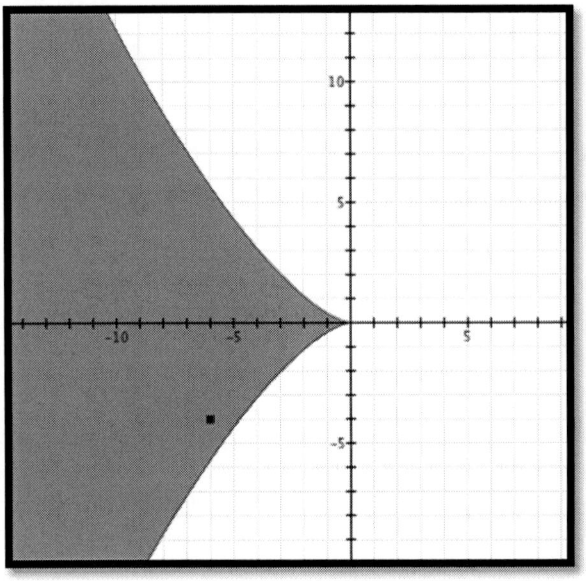

Figure 2.18. The point (-6, -4) belongs to the region with three real roots.

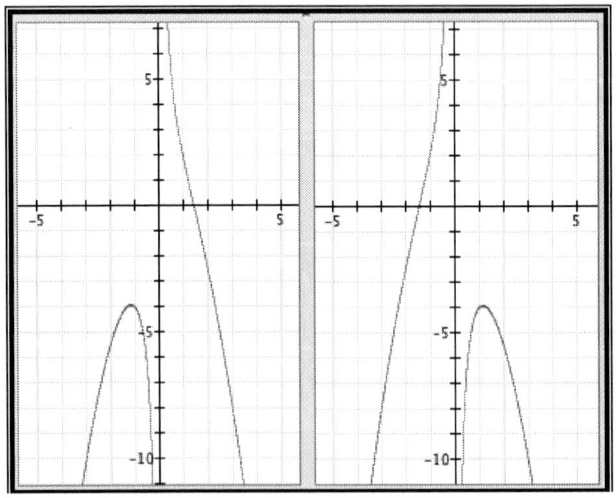

Figure 2.19. Locus of equation (20) in the plane (x, p): left—$q < 0$; right—$q > 0$.

Exploration 20. For different (positive and negative) values of parameter q draw the locus of equation (20) in the plane (x, p). Explore the dependence of the locus on parameter q.

Reflections. One can see (Figure 2.19) that, given q, equation (20) has real solutions for all values of parameter p. There may be one, two, or three different roots. To find the only local maximum that the locus of equation (20) has in the plane (x, q), consider the function $p(x) = -x^2 - \frac{q}{x}$ with parameter q. Setting $p'(x) = 0$ yields $-2x + \frac{q}{x^2} = 0$ whence $x = \sqrt[3]{\frac{q}{2}}$. Note that the inequality $q > 0$ implies the inequality $x > 0$, and the inequality $q < 0$ implies the inequality $x < 0$. Therefore, in either case, the local maximum of the function $p(x)$ is equal to

$$p(\sqrt[3]{\frac{q}{2}}) = -\sqrt[3]{(\frac{q}{2})^2} - \frac{q}{\sqrt[3]{\frac{q}{2}}} = -3\sqrt[3]{\frac{q^2}{4}}.$$

Finally, one can conclude that for all values of $q \neq 0$ equation (20) has 1) three different real roots for all $p < -3\sqrt[3]{\frac{q^2}{4}}$, 2) two different real roots for $p = -3\sqrt[3]{\frac{q^2}{4}}$, and 3) one real root for all $p > -3\sqrt[3]{\frac{q^2}{4}}$. Just like in the case of Exploration 19, the plane (q, p) can be partitioned into regions corresponding to different numbers of real roots of equation (20).

Remark 8. One can find the local maximum of the function $p(x) = -x^2 - \frac{q}{x}$ by using the Arithmetic Mean—Geometric Mean inequality (Chapter 3). To this end, in the case $q > 0$ and $x > 0$ one can write

$$-p(x) = x^2 + \frac{q}{x} = x^2 + \frac{q}{2x} + \frac{q}{2x} \geq 3\sqrt[3]{(x^2)(\frac{q}{2x})(\frac{q}{2x})} = 3\sqrt[3]{\frac{q^2}{4}},$$

that is, $-p(x) \geq 3\sqrt[3]{\frac{q^2}{4}}$, whence

$$p(x) \leq -3\sqrt[3]{\frac{q^2}{4}} \qquad (21)$$

when $q < 0$ and $x < 0$, both x^2 and $\frac{q}{x}$ are positive, thus, once again, the Arithmetic Mean—Geometric Mean inequality yields inequality (21).

Exploration 21. Consider equation (20) when $p > 0$. Determine the values of parameter q for which x^*—the only real root of the equation (as the right part of Figure 2.17 demonstrates) satisfies the inequalities

1) $x^* > q$;
2) $x^* < q$.

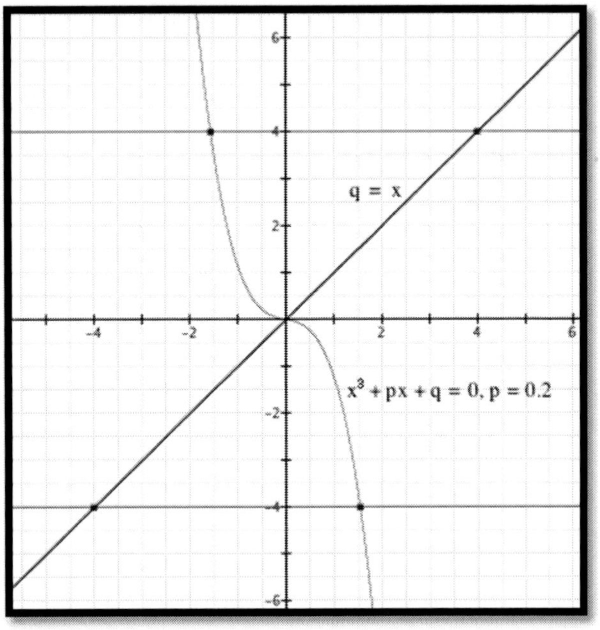

Figure 2.20. Illustration for Exploration 21.

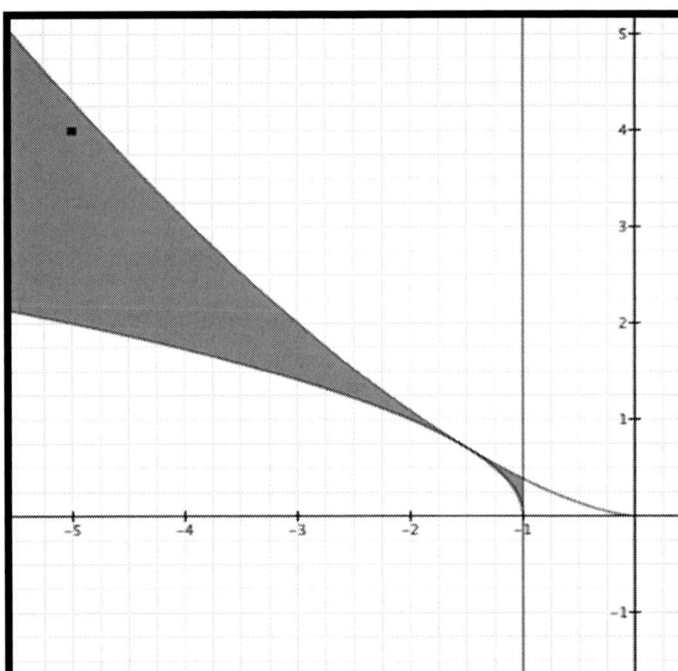

Figure 2.21. The locus of inequalities (22) in the plane (p, q) includes the point $(-5,4)$.

Reflections. As shown in Figure 2.20, which pictures the locus of equation (20) constructed in the plane (x, q) in the case $p > 0$, the line $q = constant$ first intersects the locus and then the line $q = x$ for $q > 0$; for $q < 0$ the order in which the lines intersect is opposite.

Therefore, when $p > 0$ the inequality $x^* > q$ holds true for $q > 0$ and the inequality $x^* < q$ holds true for $q < 0$.

Exploration 22. It is known that equation (20) has three real roots, $x_1 < x_2 < x_3$. In the plane (p, q) construct a region the points of which satisfy the inequalities

$$x_1 < x_2 < x_3 < q \tag{22}$$

Reflections. Consider the locus of equation (20) constructed in the plane (x, q) and pictured in Figure 2.17 (left)—the case when three real roots exist. Thus the region sought belongs to the half-plane $p < 0$. As discussed in Exploration 12, one has to use the bisector, $q = x$, of the first and the third coordinate angle in order to compare values that belong to different coordinate axes. In order to find the point of intersection of the bisector with the locus, one has to substitute x for q in equation (20) and find the positive root x^* of the equation $x^3 + px + x = 0$. Dividing both sides of the last equation by x yields $x^2 + p + 1 = 0$ whence $x^* = \sqrt{-(p+1)}$, where $p+1<0$ or $p<-1$. Using the results of Exploration 19, one can conclude that the inequalities

$$\sqrt{-(p+1)} < q < -\frac{2p}{3}\sqrt{-\frac{p}{3}} \tag{23}$$

are necessary and sufficient for the roots of equation (20) to satisfy inequalities (22).

The shaded region in Figure 2.21 shows the locus of inequalities (22). One can check to see that when $(p,q)=(-5,4)$, equation (20) turns into the equation $x^3 - 5x + 4 = 0$, which has three real roots satisfying inequalities (22). Indeed, factoring the trinomial

$$x^3 - 5x + 4 = x^3 - 1 - 5x + 5 = (x-1)(x^2 + x + 1) - 5(x-1)$$
$$= (x-1)(x^2 + x - 4) = (x-1)(x - \frac{-1+\sqrt{17}}{2})(x - \frac{-1-\sqrt{17}}{2})$$

yields the roots $1, \frac{-1+\sqrt{17}}{2}, \frac{-1-\sqrt{17}}{2}$ satisfying the inequalities

$$\frac{-1-\sqrt{17}}{2} < 1 < \frac{-1+\sqrt{17}}{2} < 4.$$

6. Systems of Simultaneous Equations with Parameters

In this section, we consider the use of geometric techniques in exploring the systems of two simultaneous equations with parameters.

Exploration 23. Partition the plane of parameters (a, b) into regions that correspond to different number of solutions of the system of equations

$$x^2 + y^2 = 2a, \quad x + y = b \tag{24}$$

Reflections. By substituting $y = b - x$, the system of equations (24) can be reduced to a single equation $x^2 + (b - x)^2 = 2a$ whose standard form is $2x^2 - 2bx + b^2 - 2a = 0$. Using the quadratic formula yields

$$x_{1,2} = \frac{b \pm \sqrt{4a - b^2}}{2}$$

Therefore, the inequality $4a - b^2 > 0$ determines the region in the plane (a, b) where the system of equations (24) has two pairs of solutions

$$(x_1, y_1) = \left(\frac{b + \sqrt{4a - b^2}}{2}, \frac{b - \sqrt{4a - b^2}}{2}\right),$$
$$(x_2, y_2) = \left(\frac{b - \sqrt{4a - b^2}}{2}, \frac{b + \sqrt{4a - b^2}}{2}\right).$$

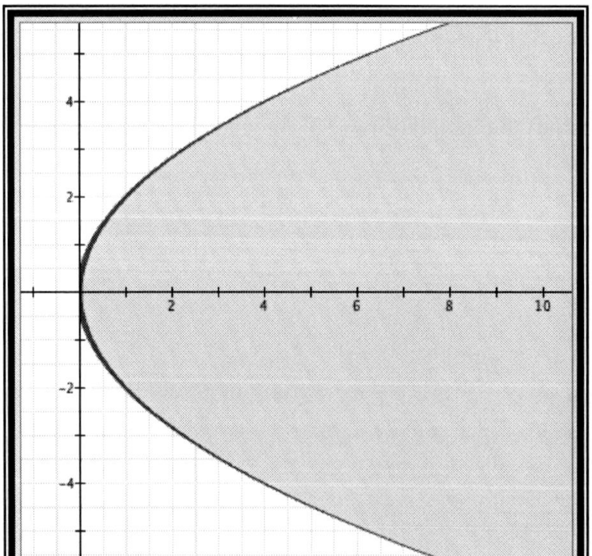

Figure 2.22. The result of Exploration 23.

When $b^2 = 4a$, there is a single solution $(x, y) = (\frac{b}{2}, \frac{b}{2})$. Finally, when $b^2 > 4a$ there are no real solutions to the system of equations (24). Figure 2.22 shows the parabola $b^2 = 4a$ that separates the plane (a, b) into two regions: with two distinct real solutions (the shaded part of the plane) and no real solutions (the rest of the plane). The parabola itself contains all pairs of parameters (a, b) that provide a single solution to the system of equations (24).

Note that the same result can be obtained geometrically in the plane (x, y) as shown in Figure 2.23. The triangle OCB is a right triangle and the Pythagorean theorem yields

$$\overline{BC} = \sqrt{b^2 - 2a}\,.$$

On the other hand,

$$\overline{BC} = \frac{\sqrt{2}}{2}b\,.$$

The equation $\sqrt{b^2 - 2a} = \frac{\sqrt{2}}{2}b$ is equivalent to the equation $b^2 = 4a$. Therefore, when $b^2 > 4a$ the straight line $x + y = b$ and the circle $x^2 + y^2 = 2a$ do not have points in common; when $b^2 < 4a$—there are two points in common; when $b^2 = 4a$—there is only one point in common.

Exploration 24. In the plane of parameters (a, b) construct a region the points of which provide the system of simultaneous equations

$$ax + by = 1, \quad x^2 + y^2 = 1 \qquad (25)$$

with solutions (x, y) such that $x > y$. Describe this region analytically in terms of inequalities between a and b.

Reflections. The first step is to solve system (25) by substitution. To this end, one can find that

$$y = \frac{1 - ax}{b} \qquad (26)$$

and, therefore,

$$x^2 + \left(\frac{1 - ax}{b}\right)^2 = 1\,.$$

The last (quadratic) equation can be transformed to the following standard form

$$(a^2 + b^2)x^2 - 2ax + 1 - b^2 = 0$$

so that, using the quadratic formula, one has

$$x_1 = \frac{a + b\sqrt{a^2 + b^2 - 1}}{a^2 + b^2},\ x_2 = \frac{a - b\sqrt{a^2 + b^2 - 1}}{a^2 + b^2}.$$

Substituting x_1 and x_2 for x in (26) yields

$$y_1 = \frac{b - a\sqrt{a^2 + b^2 - 1}}{a^2 + b^2},\ y_2 = \frac{b + a\sqrt{a^2 + b^2 - 1}}{a^2 + b^2}.$$

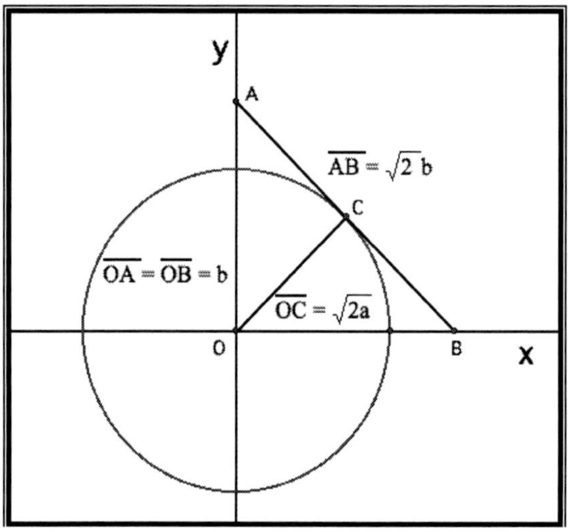

Figure 2.23. A geometric representation of equations (24) in the plane (x, y).

In that way, assuming $a^2 + b^2 \geq 1$, simultaneous equations (25) have two pairs of solutions

$$(x_1, y_1) = (\frac{a + b\sqrt{a^2 + b^2 - 1}}{a^2 + b^2}, \frac{b - a\sqrt{a^2 + b^2 - 1}}{a^2 + b^2}) \tag{27}$$

and

$$(x_2, y_2) = (\frac{a - b\sqrt{a^2 + b^2 - 1}}{a^2 + b^2}, \frac{b + a\sqrt{a^2 + b^2 - 1}}{a^2 + b^2}) \tag{28}$$

The second step is to construct the region in the plane (a, b) where the simultaneous inequalities

$$\frac{a + b\sqrt{a^2 + b^2 - 1}}{a^2 + b^2} > \frac{b - a\sqrt{a^2 + b^2 - 1}}{a^2 + b^2},$$

$$\frac{a - b\sqrt{a^2 + b^2 - 1}}{a^2 + b^2} > \frac{b + a\sqrt{a^2 + b^2 - 1}}{a^2 + b^2} \tag{29}$$

hold true. This construction can be supported by the use of the *GC*.

Because $a^2 + b^2 > 0$, inequalities (29) can be simplified to the equivalent form

$$a+b\sqrt{a^2+b^2-1} > b-a\sqrt{a^2+b^2-1},\ a-b\sqrt{a^2+b^2-1} > b+a\sqrt{a^2+b^2-1}$$

or

$$(a+b)\sqrt{a^2+b^2-1}+(a-b)>0, (a+b)\sqrt{a^2+b^2-1}-(a-b)<0 \qquad (30)$$

The region in the plane (*a*, *b*) defined by inequalities (30) coincides with the region defined by a single inequality

$$\sqrt{(a+b)\sqrt{a^2+b^2-1}+(a-b)}\sqrt{-(a+b)\sqrt{a^2+b^2-1}+(a-b)}>0$$

the locus of which is shown in Figure 2.24 (as usual, in the context of the *GC* we use custom variables *x* and *y* in place of *a* and *b*, respectively).

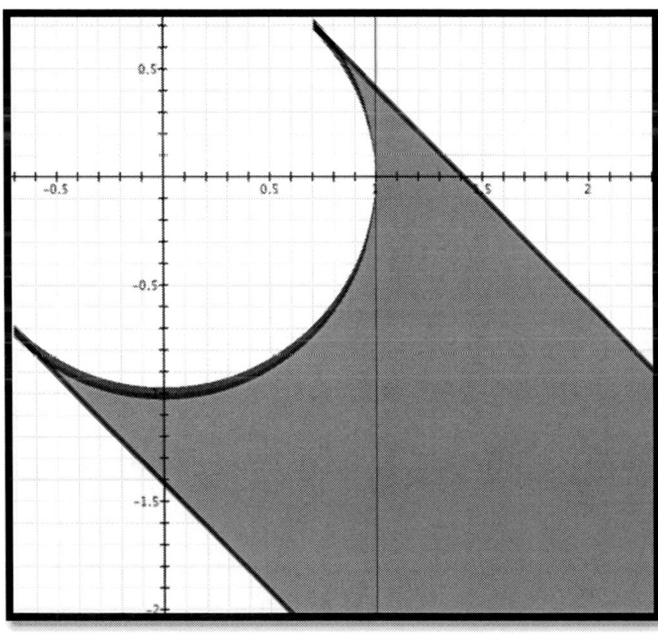

Figure 2.24. The locus of inequalities (30).

The third (and the last) step is to describe the locus of inequalities (30) in terms of parameters *a* and *b*. To this end, one has to solve the equations

$$(a+b)\sqrt{a^2+b^2-1}+(a-b)=0 \qquad (31)$$

and

$$(a+b)\sqrt{a^2+b^2-1}-(a-b)=0 \qquad (32)$$

Rewriting equation (31) in the form $(a+b)\sqrt{a^2+b^2-1}=b-a$ and squaring both sides of the last equation yields

$$(a^2+2ab+b^2)(a^2+b^2-1)=a^2-2ab+b^2,$$

whence, setting $a^2+b^2=u$ and $2ab=v$,

$$(u+v)(u-1)=u-v$$

The last equation can be simplified to the form

$$u(u+v-2)=0$$

whence $u = 0$ and $u + v = 2$. The first case is obviously extraneous as neither a nor b is equal to zero. Therefore, taking into account that $u+v=a^2+b^2+2ab=(a+b)^2$, one has $(a+b)^2=2$, whence

$$a+b=\sqrt{2} \text{ or } a+b=-\sqrt{2}.$$

It follows from Exploration 23 that the line $a+b=\sqrt{2}$ is tangent to the circle $a^2+b^2=1$ at the point $(\frac{1}{\sqrt{2}},\frac{1}{\sqrt{2}})$. Similarly, the line $a+b=-\sqrt{2}$ is tangent to the circle $a^2+b^2=1$ at the point $(-\frac{1}{\sqrt{2}},-\frac{1}{\sqrt{2}})$. Moreover, solving equation (32) would yield same results. These findings demonstrate that the boundaries of the locus of inequalities (30) coincide with the circle $a^2+b^2=1$ and the lines $a+b=\pm\sqrt{2}$.

Finally, the locus can be described through the following inequalities:

When $a \geq 1$ we have $-\sqrt{2}-a<b<\sqrt{2}-a$; when $\frac{1}{\sqrt{2}}<a<1$ we have

$$-\sqrt{2}-a<b<-\sqrt{1-a^2}$$

or

$$\sqrt{1-a^2}<b<\sqrt{2}-a;$$

when $-\frac{1}{\sqrt{2}} < a < \frac{1}{\sqrt{2}}$ we have $-\sqrt{2} - a < b < -\sqrt{1-a^2}$.

The task of the construction of the borders of the locus of inequalities (30) is included in the activity set of this chapter. The results of Exploration 24, including formulas (27)-(28), will be used later in Chapter 7.

7. Activity Set

1. For which values of b and c does equation (10) have both roots smaller than the number 1? In the plane (c, b), construct the locus of the region found.
2. For which values of b and c does equation (10) have roots separated by the number 1? In the plane (c, b), construct the locus of the region found.
3. For which values of b and c does equation (10) have both roots greater than the number 1? In the plane (c, b), construct the locus of the region found.
4. Solve the system of equations
   $x + y = a, \quad xy = b$.
   and partition the plane of parameters (a, b) into the regions with different number of solutions.
5. Solve the system of equations
   $x^2 + y^2 = 2a - 1, \quad x + y = a$.
6. Solve the system of equations
   $x^2 + y^2 = 2a, \quad x + y = b$.
7. Solve the system of equations
   $x^2 + y^2 = a^2, \quad |x| + |y| = 1$.
8. Solve the system of equations
   $x^2 + y^2 = a^2, \quad |x| + |y| = b$.
9. It is known that the equation $x^3 + px + 1 = 0$ has three real roots. Using the locus approach, demonstrate that there exists $\varepsilon > 0$ such that for all $r \in (p - \varepsilon, p + \varepsilon)$ the equation $x^3 + rx + 1 = 0$ has three roots also.
10. It is known that the equation $x^3 + px + q = 0$ has three real roots, $x_1 < x_2 < x_3$. In the plane (p, q) construct a region the points of which provide the following inequalities $x_1 < x_2 < q < x_3$, $x_1 < q < x_2 < x_3$, $q < x_1 < x_2 < x_3$.
11. It is known that the equation $x^3 + px + q = 0$ has three real roots, $x_1 < x_2 < x_3$. In the plane (p, q) construct a region the points of which provide the following inequalities $x_1 < x_2 < p < x_3$, $x_1 < p < x_2 < x_3$, $p < x_1 < x_2 < x_3$.
12. Using the results of Exploration 24, define the boundaries (an arc and two rays) of the locus of inequalities (30) through two-variable inequalities.
13. Partition the plane of parameters (q, p) into regions corresponding to different number of real roots of equation (20). Then choose a point (q, p) that provides equation (2) with one integer root and solve the equation by factoring.

14. Let x_1 and x_2 be real roots of the equation $ax^2 + x - 1 = 0$. For what values of parameter a do the inequalities $x_1 < x_2 < a$ (alternatively, $a < x_1 < x_2$ or $x_1 < a < x_2$) hold true?
15. Construct the locus of the equation $ax^2 + 2x - 1 = 0$ in the plane (x, a). Use the quadratic formula to find the roots of the equation. How are the graphs of the two roots and the locus of the equation related?
16. Construct the loci of equations $x^2 - bx - 1 = 0$ and $x^2 + bx - 1 = 0$. Show experimentally that the locus of the latter equation if a rotation about the origin of the locus of the former equation. Find the angle of rotation using one of the trigonometric functions. Confirm your findings by substituting the angle of rotation for B in the formulas $x_1 = x\cos B - y\sin B$, $y_1 = x\sin B + y\cos B$; that is, using the resulting relationships between (x, y) and (x_1, y_1), translate one quadratic equation into the other.
17. Construct the loci of the equations $ax^2 + 2x - 1 = 0$ and $ax^2 + 2x + 1 = 0$ in the plane (x, a). Show experimentally that the locus of the latter equation is a rotation about the origin of the locus of the former equation. Find the angle of rotation using one of the trigonometric functions. Confirm your findings by substituting the angle of rotation for B in the formulas $x_1 = x\cos B - y\sin B$, $y_1 = x\sin B + y\cos B$; that is, using the resulting relationships between (x, y) and (x_1, y_1), translate one quadratic equation into the other.
18. Show that the sequence of points (x_n, y_n) satisfying the system of difference equations

$$x_n + y_{n-1} = a, \quad x_{n-1}y_n = b, \quad x_1 = y_1 = 1$$

converges to the point $\left(\frac{a + \sqrt{a^2 - 4ab}}{2}, \frac{a - \sqrt{a^2 - 4ab}}{2}\right)$ when $a^2 - 4ab \geq 0$.

Chapter 3

INEQUALITIES AND SPREADSHEET MODELING

He who seeks for methods without having a definite problem in mind seeks for the most part in vain.

— David Hilbert[1]

(Oxford Dictionary of Scientific Quotations, 2005, p. 280)

1. INTRODUCTION

Consider the "problem" of finding the sum 36 + 27. Let us pretend that this is a difficult task because we do not know well how to add numbers except in some simple cases. Let 36 + 27 = x. Can we somehow *estimate* the value of x? Ironically, the estimation of integers is a more basic skill than the skill of adding integers. Indeed, one can grasp the concepts "more" and "less" earlier than the concept "equal," as the recognition of difference based on perception, comes earlier than the recognition of similarity based on operation (Piaget, 1954). So, we know that 36 is greater than 30 and 27 is greater than 20. Therefore, the sum 36 + 27 is greater than the sum 30 + 20. The latter sum is easier to find than the former sum. Thus, 36 + 27 is greater than 50. By the same token, as 36 is smaller than 40 and 27 is smaller than 30, the sum 36 + 27 is smaller than 70. Symbolically, our findings can be expressed in the form of the inequalities

$$50 < x < 70 \tag{1}$$

In other words, our sum, x, can be any number greater than 50 and smaller than 70.

Can this estimation be improved by narrowing the range of numbers to which x belongs? Well, let us assume that we know how to add numbers with the last digit 5. Noting that 36 is greater than 35 and 27 is greater than 25, the sum 36 + 27 is greater than 35 + 25 = 60 and, thereby, inequality (1) can be improved as follows

$$60 < x < 70 \tag{2}$$

[1] David Hilbert (1862-1943)—a German mathematician who is best known for putting forward in 1990 a list of 23 unsolved problems thus, to a large extent, setting the course of the 20th century mathematics.

There are nine integers that satisfy inequalities (2). Yet, four of them can be excluded as possible values of x if one recognizes that the sum of two numbers of different parity, 36 + 27, is an odd number. Furthermore, the sum cannot have 7 as the last digit as, whatever the sum 6 + 7 is, it cannot end with the digit 7. Therefore $x \in \{61, 63, 65, 69\}$.

So, in our attempt to find the sum 36 + 27 we made a significant progress by narrowing the search to four numbers only. This progress was made possible by our *use of inequalities*.

Inequalities are among the most useful mathematical tools. When solving a problem, these tools enable one to move from not knowing the answer at all to knowing at least something about it through estimation. This chapter, in accord with standards for teaching and recommendations for teachers in North America, will address the issue of preparing teachers in the use of inequalities as problem-solving tools in the context of constructing computationally efficient spreadsheet-based environments.

The Principles and Standards for School Mathematics (National Council of Teachers of Mathematics, 2000) recommend that in grades 9-12 all students should "understand the meaning of equivalent forms of expressions, equations, inequalities and relations; write equivalent forms of equations, inequalities, and systems of equations and solve them with fluency … using technology in all cases" (p. 296). In addition, "instructional programs … should enable all students to select and use various types of reasoning and methods of proof" (p. 342). These ambitious expectations for high school mathematics curricula and teaching undoubtedly raise the level of professional standards for the teachers from their current position.

Indeed, the recommendations for the preparation of teachers in algebra and number theory provided by the Conference Board of the Mathematical Sciences (2001) include the need for teachers to understand "the ways that basic ideas of number theory and algebraic structures underlie rules for operations on expressions, equations, and inequalities" (p. 40) and the importance of courses within which teachers "could examine the crucial role of algebra in use of computer tools like spreadsheets and the ways that [technology] might be useful in exploring algebraic ideas" (p. 41). These recommendations point at the important role that training in the joint use of inequalities and digital technology should play in the preparation of teachers.

In this chapter it will be shown how a spreadsheet can be used as a milieu for teachers' training in the use of inequalities and associated proof techniques. In some cases, a spreadsheet will be used as a generator of problems on computational efficiency leading to the use of inequalities. In other cases, a context within which computational environments were created will be extended to allow for inequalities to be used as problem-solving tools. Thus, we will shift the focus from a traditional pedagogy of utilizing technology for *solving* inequalities to *using* inequalities as problem-solving tools in computing applications.

2. SPREADSHEET MODELING OF LINEAR EQUATIONS

Many problems found across pre-college mathematics curriculum can be reduced to linear equations in two variables. The case of Diophantine equations[2] is of particular importance for spreadsheet-enabled mathematics—using a spreadsheet one can generate solutions to such equations by computing values of linear combinations of pairs of whole numbers and comparing these values to the right-hand side of equation in question. As an example, consider

Problem 1. A bike store's owner sells only bicycles and tricycles. One day she asked her clerk to count how many vehicles there were in the store. Instead of counting vehicles, the clerk counted wheels and reported the total of 18 wheels. How many bicycles and tricycles might there have been?

In designing a spreadsheet-based environment for numerical modeling of the bike store problem the following problem-solving situation arises[3]: *Given the total number of wheels among vehicles, determine the maximum number of each type of vehicle that might have been in the store*. In a de-contextualized form, the problem is to find the greatest values of variables x and y which satisfy the equation

$$ax + by = n \qquad (3)$$

with whole number coefficients a and b ($n = 18$, $a = 2$, and $b = 3$ in the case of Problem 1). Knowing such values of x and y (that is, the largest total for each type of vehicle) enables one to generate solutions within electronic charts that do not include unnecessary computations. It is through resolving such a computational problem that pedagogically useful activities involving appropriate use of inequalities in the context of proving can come into play.

To begin, note that rough upper estimates for an x-range and y-range are quite apparent: $x \le n$ and $y \le n$. In mathematics, however, even an apparent statement requires formal demonstration. Therefore, these simple inequalities can be used as a springboard into content-specific proof techniques. One such technique is based on reasoning known as *proof by contradiction*—where, for the sake of argument, one makes an assumption contrary to what has to be proved, arrive at an absurd result, and then conclude that the original assumption must have been wrong, since it led to this result. This type of argument (sometimes referred to as *reduction to an absurdity*) makes use of the so-called *law of excluded middle*—a statement that cannot be false, must then be true.

Proposition 1. The inequality $x \le n$ holds true.

Proof. Suppose that, on the contrary, $x > n$. Then for any $y \ge 0$ it follows that $n = ax + by > an \ge n$. This contradiction (i.e., the false inequality $n > n$) suggests that x cannot be greater than n, thus $x \le n$. This completes the proof.

---

2 It is customary to apply the term Diophantine equation to any indeterminate equation in one or more variables with integer coefficients that has to be solved in integers. The term derives from the name of the Greek mathematician Diophantus who lived in Alexandria in the 3rd century A.D.

3 In the words of Euler, "Since the fabric of the world is the most perfect and was established by the wisest Creator, nothing happens in this world in which some reason of maximum or minimum would not come to light" (cited in Pólya, 1954, p. 121). Reasoning along these lines will continue being a focus of the discussion throughout this chapter.

It may be helpful to repeat this proof using the bike store context so that each mathematical sentence can be supported by a situational referent. To this end, suppose that the number of bicycles, x, is greater that the total number of wheels counted, n. Then, for any number of tricycles, y, the total number of wheels is greater than twice the number of bicycles. The latter, in turn, is greater that the number of bicycles[4]. But the number of bicycles, x, was assumed to be greater than the total number of wheels counted, n. In that way, we arrived to a contradictory conclusion: the total number of wheels in the store is greater than the total number of wheels in this store. Therefore, the original assumption about the number of bicycles in the store was wrong; i.e., the number of bicycles in the store is smaller than the total number of wheels counted.

Being intuitively apparent in contextual terms and thus didactically useful, the above upper estimates of ranges for variables x and y in equation (1) can be significantly improved. The need for such an improvement can be computationally driven. Indeed, by generating solutions to Problem 1 within an 18×18 table one can see that the number of cells within which computations have to be carried out may be reduced nine-fold. Using a combination of formal and contextual arguments, one can prove (by contradiction)

Proposition 2. The inequality $x \le \frac{n}{a}$ holds true.

Proof. Suppose, on the contrary, $x > \frac{n}{a}$ for any $y > 0$. This assumption results in the following contradictory conclusion: $n = ax + by > a \cdot \frac{n}{a} = n$. This completes the proof.

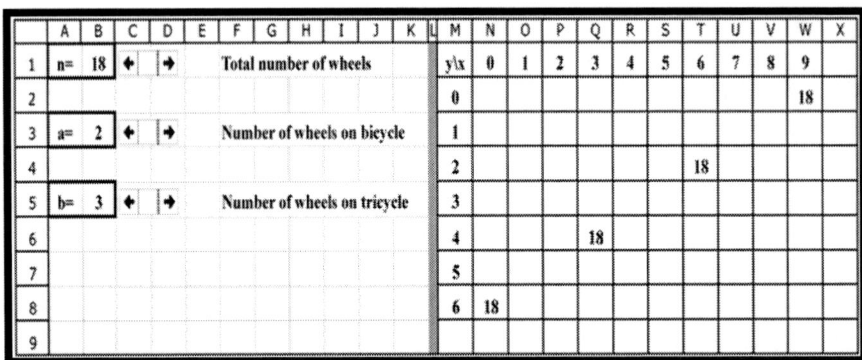

Figure 3.1. Four solutions displayed.

Furthermore, taking into account that x is an integer variable and using the function INT(x) which rounds x down to the nearest integer, yield an even stronger inequality $x \le$ INT(n/a). Similarly, the inequality $y \le$ INT(n/b) can be used as an upper estimate for the variable y. As an application of the last two inequalities to modeling equation (3), spreadsheets pictured in Figures 3.1 and 3.2 generate computationally efficient x- and y-ranges for different values of n. The knowledge of lower estimates for variables x and y can be used to further improve computational efficiency of the environment in question. For

4 By using a non-strict inequality $an \ge n$ above we allowed for the case $a = 1$.

example, the inequality $y \geq k$, where k is a positive integer smaller than n, enables for the following improvement of the x-range found earlier.

Proposition 3. When $y \geq k$, the inequality $x \leq \frac{n-bk}{a}$ holds true.

	A	B	C	D	E	F	G	H	I	J	K	L	M	N	O	P	Q	R	S	T	U	V	W	X	Y	Z
1	n=	24	←	→		Total number of wheels							y\x	0	1	2	3	4	5	6	7	8	9	10	11	12
2													0													24
3	a=	2	←	→		Number of wheels on bicycle							1													
4													2										24			
5	b=	3	←	→		Number of wheels on tricycle							3													
6													4							24						
7													5													
8													6				24									
9													7													
10													8	24												

Figure 3.2. Five solutions displayed.

Proof. The inequalities $y \geq k$ and $b > 0$ imply the inequality $by \geq bk$ whence $n - by \leq n - bk$. Therefore, it follows from equation (3) that $x = \frac{n-by}{a} \leq \frac{n-bk}{a}$. This completes the proof.

Remark 1. Noting that $\frac{n-bk}{a} < \frac{n}{a}$ shows that Proposition 3 provides an improved upper bound for x in comparison with Proposition 2. Finally, taking into account that x is an integer variable results in a stronger inequality, $x \leq INT(\frac{n-bk}{a})$.

	A	B	C	D	E	F	G	H	I	J	K	L	M	N	O	P	Q	R	S	T	U
1	n=	18	←	→		Total number of wheels							y\x	0	1	2	3	4	5	6	
2													2							18	
3	a=	2	←	→		Number of wheels on bicycle							3								
4													4				18				
5	b=	3	←	→		Number of wheels on tricycle							5								
6													6	18							
7	Lower bound for x				0	←	→														
8	Lower bound for y				2	←	→														

Figure 3.3. The larger y_0, the smaller x-range.

Spreadsheets in which information about a lower estimate for one variable is used to improve an upper estimate for another variable are pictured in Figures 3.3 and 3.4 (see Appendix for programming details). They show that the larger the lower bound for the variables *y* and *x*, respectively, the smaller the upper bound for those variables. This computationally driven statement can be interpreted in the following contextual terms: the more vehicles of one type the store has, the fewer vehicles of another type are there. In that way, the meaning of inequalities can be communicated to the learners of mathematics in three different ways: contextually, computationally, and mathematically. A similar didactic approach will be used in the next section allowing for new inequalities and associated proof techniques to be discussed.

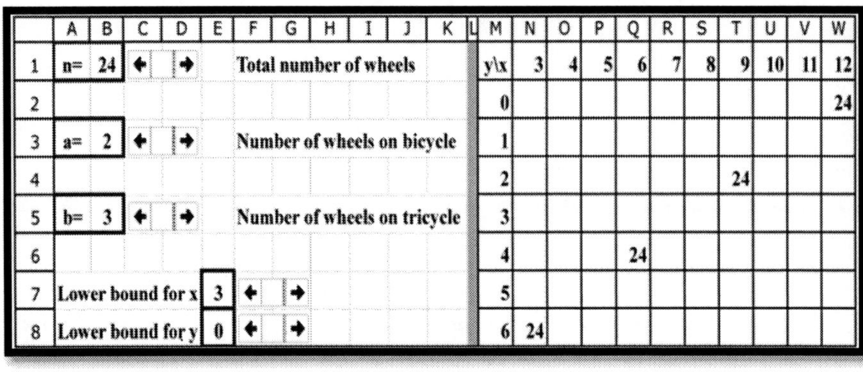

Figure 3.4.The larger x_0, the smaller *y*-range.

3. Inequalities as Tools in Modeling Non-linear Problems

Interesting activities in the use of inequalities and proof techniques can be carried out in the context of spreadsheet modeling of non-linear equations that represent unit fractions as the sum of two or more like fractions. There are problems both within and outside mathematics in which such representations are important. Consider

Problem 2. Small square-shaped desks each of which can seat exactly four students are put together to form a large rectangular table that can seat as many students as the number of the desks used. How many ways can this be done?

Setting *x* and *y* to be integer dimensions of a rectangle (alternatively, the number of desks used to form each of the two adjacent sides), this problem can be reduced to a non-linear algebraic equation $xy = 2x + 2y$. Dividing both sides of the last equation by $2xy$ yields

$$\frac{1}{2} = \frac{1}{x} + \frac{1}{y} \qquad (4)$$

Equation (4) represents a mathematical model of Problem 2. Put another way, equation (4) defines all ways to represent a unit fraction 1/2 as a sum of two like fractions. Below, however, a more general equation will be explored

$$\frac{1}{n}=\frac{1}{x}+\frac{1}{y} \tag{5}$$

from which, in particular, solution to Problem 2 will be found.

Equation (5) can serve as a mathematical model for the family of rectangles with integer sides and area, numerically, n times as much as its semi-perimeter. Indeed, multiplying both sides of equation (5) by nxy yields $xy = n(x + y)$ where xy and $x + y$ are, respectively, area and semi-perimeter of a rectangle with integer dimensions x and y. Furthermore, equation (5) implies the inequality $x > n$; indeed, in the case $x \le n$ the chain of inequalities

$$\frac{1}{n}=\frac{1}{x}+\frac{1}{y}\ge\frac{1}{n}+\frac{1}{y}>\frac{1}{n}$$

leads to the contradiction $\frac{1}{n}>\frac{1}{n}$. Finally, due to symmetric nature of equation (5), we assume $n<x\le y$.

As in the case of equation (3), in designing a spreadsheet-based environment for modeling solutions to equation (5) the following context-bounded inquiry into its structure can be raised:

Problem 3. Given n (the ratio of area to semi-perimeter of a rectangle), determine the largest values for each of the variables x and y (dimensions of the rectangle) satisfying equation (5).

Unlike equation (3), in which rough (upper) estimates for x and y-ranges could be found almost intuitively, equation (5), where x, y, and n are related to each other in a non-linear way, does not allow for such an intuitive approach. A first step in finding upper estimates for x and y in equation (5) could be to make use of

Proposition 4. The identity

$$\frac{1}{n}=\frac{1}{n+1}+\frac{1}{n(n+1)} \tag{6}$$

holds true for all $n = 1, 2, 3, \dots$.

Proof. Multiplying both sides of (6) by $n(n + 1)$ yields $n + 1 = n + 1$, an obviously true identity.

Remark 2. Another way of confirming identity (6) is by evaluating the difference of the reciprocals of two consecutive counting numbers

$$\frac{1}{n}-\frac{1}{n+1}=\frac{n+1-n}{n(n+1)}=\frac{1}{n(n+1)}.$$

This approach was not based on our knowledge of identity (6); rather, it was discovered using algebraic transformations. The two cases show two distinct ways of formal demonstration (proving) in mathematics: moving from a conjecture to a true statement and developing a true statement through a correct application of rules of mathematics.

Identity (6), sometimes attributed to Fibonacci (Hoffman, 1998), represents a unit fraction $1/n$ as the sum of the largest and the smallest possible unit fractions (or, alternatively, as will be shown below, through the dimensions of the rectangle with the largest perimeter) satisfying equation (5). This observation brings about preliminary upper and lower estimates for x and y in the form of the inequalities

$n+1 \le x \le n(n+1)$ and $n+1 \le y \le n(n+1)$.

Having experience with modeling a linear problem in the form of equation (3), one may wonder: Could these inequalities be improved? With this in mind, note that if $n > 1$, then $n(n+1) > 2n$. Indeed, dividing both sides of the last inequality by n yields $n+1 > 2$, a true inequality. Now one can show that at most one denominator (or, alternatively, rectangle's side) may be greater than $2n$. Invoking proof by contradiction, that is, assuming $y \ge x > 2n$, one can arrive to the following absurd conclusion

$$\frac{1}{n} = \frac{1}{x} + \frac{1}{y} > \frac{1}{2n} + \frac{1}{2n} = \frac{1}{n}.$$

Therefore, if $x \le y$, then $n+1 \le x \le 2n$ and $n+1 \le y \le n(n+1)$. Likewise, at most one fraction in the right-hand side of equation (5) may have a denominator smaller than $2n$ (or, alternatively, only one side of a rectangle may be smaller than $2n$). In this way, it appears that the inequalities

$$n+1 \le x \le 2n \le y \le n(n+1), \qquad (7)$$

when compared to those previously found, improve the upper and lower bounds for, respectively, x and y, and, therefore, enable for the design of a more computationally efficient spreadsheet-based environment for modeling solutions to equation (5).

The spreadsheet that incorporates inequalities (7) in the case $n = 6$ is pictured in Figure 3.5 (*D2:I2*—the x-range; *C3:C33*—the y-range). The results of computations suggest that there are five ways to partition 1/6 into the sum of two unit fractions:

1/6 = 1/7 + 1/42, 1/6 = 1/8 + 1/24, 1/6 = 1/9 + 1/18,
1/6 = 1/10 + 1/15, 1/6 = 1/12 + 1/12.

The same spreadsheet can show that there are only two ways to partition 1/2 into the sum of two unit fractions: 1/2 = 1/3 + 1/6 and 1/2 = 1/4 + 1/4. These two equations bring about the answer to Problem 2: there are only two ways to arrange the desks in the form of rectangle that seats as many people as there are desks: a 4×4 square (a special case of rectangle) and a 3×6 rectangle.

By modeling solutions to equation (5) for different values of n one can observe the following pattern: the solution associated with identity (6) is separated from other solutions by a wide gap on a spreadsheet template. Indeed, as Figure 3.5 shows, in the case $n = 6$ this gap spans from $y = 24$ to $y = 42$; that is, about 40% of the template generates no solution. This observation enables one to further improve computational efficiency of the environment. To this end, note that, in general, because identity (6) provides a solution to equation (5) for any integer n, an upper estimate for y can be improved in comparison with the inequality $y \le n(n + 1)$.

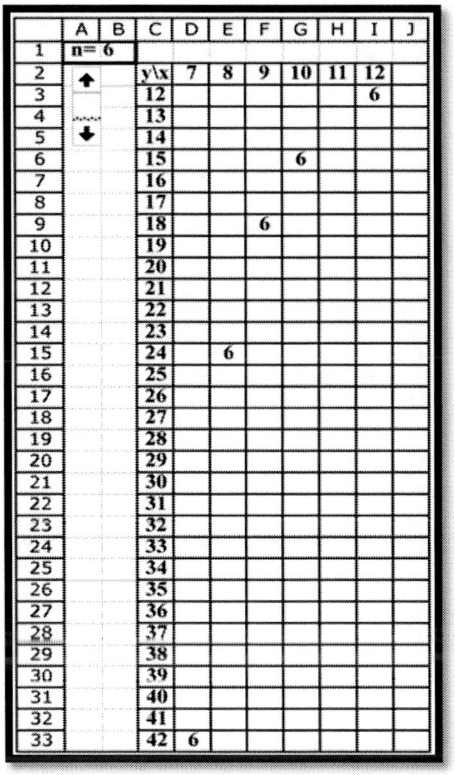

	A	B	C	D	E	F	G	H	I	J
1	n= 6									
2			y\x	7	8	9	10	11	12	
3			12						6	
4			13							
5			14							
6			15				6			
7			16							
8			17							
9			18			6				
10			19							
11			20							
12			21							
13			22							
14			23							
15			24		6					
16			25							
17			26							
18			27							
19			28							
20			29							
21			30							
22			31							
23			32							
24			33							
25			34							
26			35							
27			36							
28			37							
29			38							
30			39							
31			40							
32			41							
33			42	6						

Figure 3.5. Using inequalities (4).

Proposition 5. Let $x \le y$ satisfy equation (5) and $x > n + 1$. Then the inequalities

$$n+2 \le x \le 2n \le y \le \frac{n(n+2)}{2} \tag{8}$$

hold true.

Proof. Let $x = n + 2$ be the second smallest value of x satisfying equation (5). Then, it follows from equation (5) that $\frac{1}{y} = \frac{1}{n} - \frac{1}{n+2} = \frac{2}{n(n+2)}$ whence $y = \frac{n(n+2)}{2}$. This completes the proof of inequalities (8).

Proposition 5 enables one to further improve computational efficiency of the spreadsheet environment. One can see that $\frac{n(n+1)}{n(n+2)/2} = 2(1 - \frac{1}{n+2}) \ge 1.5$, thus the old y-range with $y_{\max} = n(n+1)$ is at least 1.5 times larger than the new (improved) one where $y_{\max} = \frac{n(n+2)}{2}$. Visually, this difference becomes apparent by comparing Figure 3.5 to

Figure 3.6, as the latter shows the spreadsheet built on inequalities (8). The next section will show how, in addition to computationally driven use of inequalities, new ideas and techniques can be introduced through the advancement of contextual arguments.

	A	B	C	D	E	F	G	H	I	J	K	L
1		Note: this table does not										
2		include the known solution						x=	7		y=	42
3	n=	6										
4		y\x	8	9	10	11	12					
5		12					6					
6		13										
7		14										
8		15			6							
9		16										
10		17										
11		18		6								
12		19										
13		20										
14		21										
15		22										
16		23										
17		24	6									

Figure 3.6. Using inequalities (5).

4. Geometric Context as a Springboard into New Uses of Inequalities

The unity of context, computing, and mathematics is a useful pedagogical triad for it enables the introduction of new techniques associated with the use of inequalities. As was mentioned above in passing, with no justification, identity (6) can be interpreted as a solution to equation (5) that corresponds to the rectangle with the largest perimeter. In what follows, it will be demonstrated how one can advance this comment to the status of being a rigorously proved mathematical proposition. It is through such advancement that new ideas and proof techniques associated with inequalities can be discussed. An important aspect of this discussion is the value of geometric roots of algebraic propositions. In what follows, the value of context in acquiring new mathematical knowledge can be further elucidated.

4.1. Finding Rectangle with the Largest Perimeter

In order to find a rectangle with the largest perimeter that satisfies the conditions of Problem 3 (i.e., equation [5]), one has to define its perimeter as a function of one of the dimensions, say y, with n being a parameter. To this end, note that, as it follows from equation (5), the other dimension is $x = \frac{ny}{y-n}$. Because $x \leq y$ with $x = y = 2n$, it follows that $2n \leq y \leq n(n + 1)$. The sum of the two dimensions, $y + x$, can be simplified as follows

$$y + x = y + \frac{ny}{y-n} = \frac{y^2}{y-n},$$

thereby, enabling one to introduce the perimeter function $P(y) = \frac{2y^2}{y-n}$ defined on the segment $[2n, n(n + 1)]$. At this point, one can be reminded of the Extreme Value Theorem (studied in a calculus course and sometimes referred to as the theorem of Weierstrass[5]), which provides a theoretical foundation for finding the largest (and the smallest) perimeter.

Extreme Value Theorem. If a function $f(x)$ is continuous on the segment $[a, b]$ then $f(x)$ has both a minimum and a maximum on $[a, b]$.

According to this theorem (which formal proof, despite apparent geometric interpretation, requires rather sophisticated lines of argumentation based on the theory of limits), the perimeter function $P(y)$ attains both the largest and the smallest values on the segment $[2n, n(n + 1)]$. In that way, following a recommendation by the Conference Board of the Mathematical Sciences (2001), an explicit connection between high school and college mathematics curricula can be established in the context of Problem 3. Furthermore, geometric context associated with equation (5) brings about the need for the use of inequalities.

The next step in our search for rectangle with the largest perimeter is to show that the function $P(y)$ of integer variable y monotonically increases so that $P(y + 1) > P(y)$ for all $y \in [2n, n(n + 1)]$. Towards this end, two auxiliary inequalities need to be established.

Lemma 1. The inequality

$$(1 + \frac{1}{y})^2 > 1 + \frac{2}{y} \tag{9}$$

holds true.

Proof. Using the formula $(a + b)^2 = a^2 + 2ab + b^2$ and noting that a square of a non-zero number is always greater than zero, one can write

---

[5] Karl Weierstrass—the outstanding German mathematician of the 19th century who, being motivated by various technical problems arising in mathematics, introduced rigor into analysis and for this work is often referred to as the "father of modern analysis."

$$(1+\frac{1}{y})^2=1+\frac{2}{y}+\frac{1}{y^2}>1+\frac{2}{y}.$$

Note that inequality (9) is a special case of a more general statement (known as Bernoulli inequality[6])

$(1+y)^r \geq 1+ry,\ y>-1, r>1$ or $r<0$,

which proof can be found, for example, in a classic, high school-oriented treatise by Korovkin (1961).

Lemma 2. If $y \geq 2n$ then

$$\frac{2}{y} \geq \frac{1}{y-n} \qquad (10)$$

Proof. Cross-multiplying the terms in inequality (10) yields $2y - 2n \geq y$ or $y \geq 2n$, the latter being the assumption of the lemma. This completes the proof.

Remark 3. One can prove inequality (10) by adding y to the both sides of the inequality $y \geq 2n$ to get $2y \geq 2n + y$ whence $2y - 2n \geq y$. Dividing both sides of the last inequality by $y(y - n)$ results in inequality (10). On the other hand, by subtracting n from both sides of the inequality $y \geq 2n$ yields $y - n \geq n$ whence $\frac{1}{n} \geq \frac{1}{y-n}$. However, comparing $\frac{1}{n}$ to $\frac{2}{y}$ shows that $\frac{1}{n} \geq \frac{2}{y}$ as well. So, by establishing that both $\frac{1}{y-n}$ and $\frac{2}{y}$ are smaller than $\frac{1}{n}$ does not allow one to prove inequality (10). This example shows that proving inequalities is not a routine but rather a skill, often developed from trial and error, in other words, from a practice of *doing* proof. Indeed, trial and error approach should not be de-emphasized but rather encouraged and promoted through the teaching of mathematics because, as Francis Bacon[7] once said, "truth comes out of error more readily than out of confusion" (cited in Bradley, 2005, p. 9).

Proposition 6. Let $P(y)=\frac{2y^2}{y-n}$ and $y \in [2n, n(n+1)]$. Then the inequality

$$\frac{P(y+1)}{P(y)}>1 \qquad (11)$$

[6] Jacob Bernoulli – a Swiss mathematician of the 17th century who posed the so-called "Basel problem" (solved by Euler) of finding the exact sum of the series $\sum_{n=1}^{\infty}\frac{1}{n^2}$ (see Chapter 1).

[7] Francis Bacon (1561-1626)—English philosopher, essayist, and statesman.

holds true on the segment $[2n, n(n+1)]$. In other words, the perimeter function $P(y)$ monotonically increases on this segment as the difference between the dimensions of the corresponding rectangle increases.

Proof. This time, another type of argument will be utilized in proving inequality (11). Based on a straightforward combination of earlier established facts, this kind of argument is commonly referred to as *direct proof.* To begin, one can start with using simple rules for operations on algebraic expressions to show that

$$\frac{P(y+1)}{P(y)} = \frac{(y+1)^2(y-n)}{y^2(y-n+1)} = \frac{y^2(1+\frac{1}{y})^2(y-n)}{y^2(\frac{y-n}{y-n}+\frac{1}{y-n})} = \frac{(1+\frac{1}{y})^2}{1+\frac{1}{y-n}} \tag{12}$$

Next, applying inequalities (9) and (10) to, respectively, the numerator and denominator of the far-right fraction in the chain of equalities (12) yields

$$\frac{(1+\frac{1}{y})^2}{1+\frac{1}{y-n}} > \frac{1+\frac{2}{y}}{1+\frac{1}{y-n}} \geq \frac{1+\frac{1}{y-n}}{1+\frac{1}{y-n}} = 1.$$

This completes the proof based on the application of two previously proved lemmas (auxiliary propositions).

In that way, because $P(y)$ is a continuous function on the segment where it monotonically increases, the largest value of $P(y)$ exists and is reached at the right endpoint $y = n(n + 1)$. Therefore,

$$P(y) \leq P(n(n+1)) = 2\frac{n^2(n+1)^2}{n^2} = 2(n+1)^2$$

for all $y \in [2n, n(n + 1)]$. In other words, identity (6), indeed, determines an integer-sided rectangle with the largest perimeter and area being n times as much as half of this perimeter.

4.2. Finding Rectangle with the Smallest Perimeter

The Extreme Value Theorem guarantees that the function $P(y)$ has both the global maximum and minimum on the segment $[2n, n(n+1)]$; in other words, in the context of Problem 3 there exists a rectangle with the smallest perimeter also. In order to find such a rectangle, one can utilize a famous analytic inequality (with profound geometric meaning) known as the Arithmetic Mean—Geometric Mean inequality:

$$\frac{u+v}{2} \ge \sqrt{uv}\text{, for } u \ge 0,\ v \ge 0 \qquad (13)$$

with equality taking place when $u=v$. Inequality (13) can be proved directly by demonstrating that the difference between its left and right-hand sides is non-negative. Indeed,

$$\frac{u+v}{2} - \sqrt{uv} = \frac{u+v-2\sqrt{uv}}{2} = \frac{(\sqrt{u})^2+(\sqrt{v})^2-2\sqrt{uv}}{2} = \frac{(\sqrt{u}-\sqrt{v})^2}{2} \ge 0.$$

The utilization of inequality (13) in estimating the function $P(y)$ cannot be carried out in a straightforward way though. One needs a specific representation of $P(y)$ based on the identity

$$\frac{y^2}{y-n} = y-n+\frac{n^2}{y-n}+2n \qquad (14)$$

which can be proved without difficulty by simplifying its right-hand side. This makes it possible to apply inequality (13) to the fraction $\dfrac{y-n+\dfrac{n^2}{y-n}}{2}$, the arithmetic mean of $y - n$ and $\dfrac{n^2}{y-n}$, and estimate the function $P(y)$ from below as follows

$$P(y) = 4\left(\frac{y-n+\dfrac{n^2}{y-n}}{2}+n\right) \ge 4\left(\sqrt{(y-n)\frac{n^2}{y-n}}+n\right) = 8n.$$

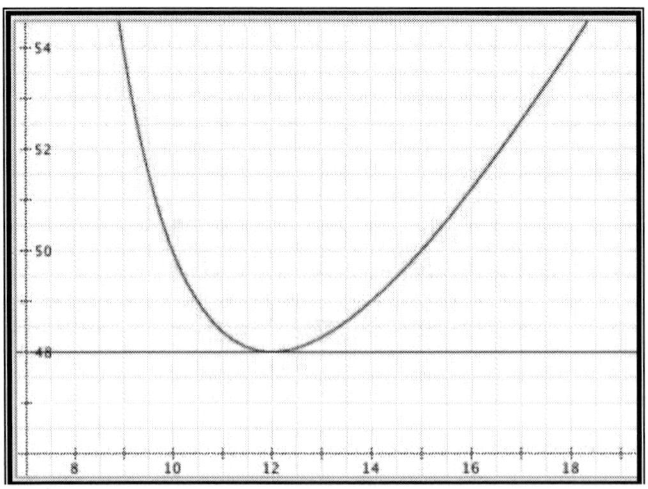

Figure 3.7. The graph of the perimeter function $P(y)$.

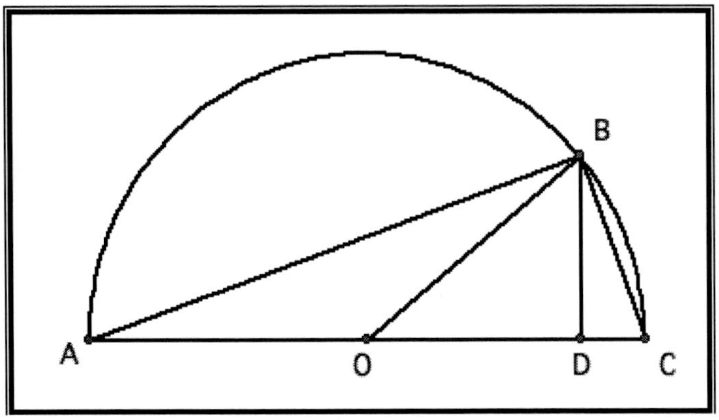

Figure 3.8. A geometric proof of inequality (13).

Note that the representation of $P(y)$ through identity (14) is due to the need of having two addends, the product of which does not depend on y. Thus $P(y) \geq 8n$ with equality taking place when $y - n = \frac{n^2}{y-n}$, that is, when $y = 2n$. In other words, rectangle with the smallest perimeter and area being n times as much as half of this perimeter is a square with side $2n$.

4.3. Geometric Proof of the Arithmetic Mean—Geometric Mean Inequality

Consider the geometric situation presented in Figure 3.8. Let O be the midpoint of diameter AC and $BD \perp AC$. $\angle ABC = 90^\circ$—an angle supported by diameter AC. $\triangle BDC$ and $\triangle ABD$ are similar right triangles, because $\angle BCD = 90^\circ - \angle DBC = \angle ABD$. The similarity implies the proportion $\frac{BD}{DC} = \frac{AD}{BD}$ whence $BD = \sqrt{AD \cdot DC}$. Note that $BO \geq BD$ and $BO = \frac{AC}{2} = \frac{AD + DC}{2}$. Setting $AD = u$ and $DC = v$ yields the inequality $\frac{u+v}{2} \geq \sqrt{uv}$ with equality taking place when $u = v$ (that is, when points D and O coincide). This completes a geometric proof of inequality (13).

4.4. Alternative Approaches to Problem 3

The solution of Problem 3 can be developed using *direct proof* methods based on geometric reasoning. One such method is to graph the function $P(y)$ and observe that $\max_{2n \leq y \leq n(n+1)} P(y) = P[n(n+1)] = 2(n+1)^2$ and $\min_{2n \leq y \leq n(n+1)} P(y) = P(2n) = 8n$ ($n = 6$ in Figure 3.7). Another method of finding the smallest perimeter is based on the fact that the only local minimum of the function $P(y)$ exists at the point $y = 2n$. Indeed, the equation

$$\frac{2y^2}{y-n} = k \tag{15}$$

has a single solution only when $k = 8n$ and, thereby, $y = 2n$. This fact, supported by the graph of the function $P(y) = \frac{2y^2}{y-n}$, can be proved algebraically by reducing equation (15) to the equation $2y^2 - ky + kn = 0$ with the roots $y_{1,2} = \frac{k \pm \sqrt{k^2 - 8kn}}{4}$. When $k^2 - 8kn = 0$, equation (15) has a double root at the point when the level line $y = k$ is tangent to the graph of the function $P(y)$. As $P(y) > 0$, so is the value of k, thus $\min_{2n \le y \le n(n+1)} P(y) = k = 8n$.

5. Transition to Three-Dimensional Modeling

Although a spreadsheet is commonly used in mathematics education as a modeling tool for problems with at most two variables, several computational methods enable its use beyond two dimensions. In the case of modeling a Diophantine equation, one such method consists of constructing level lines of integer values. As an illustration, consider the following problem that leads to a three-dimensional analogue of equation (5).

Problem 4. Find the total number of right rectangular prisms of integer sides whose volume, numerically, is n *times as much as the half of its surface area.*

Similar to Problem 3, setting x, y, and z to be integer dimensions of a right rectangular prism results in the equation $xyz = n(xy + xz + yz)$. Dividing both sides of the last equation by $nxyz$ yields

$$\frac{1}{n} = \frac{1}{x} + \frac{1}{y} + \frac{1}{z} \tag{16}$$

Equation (16), representing a unit fraction as a sum of three like fractions, is a three-dimensional analogue of equation (5). Rewriting equation (16) in the form

$$z = \frac{nxy}{xy - n(x+y)} \tag{17}$$

makes it possible to use the two-dimensional computational capacity of a spreadsheet in numerical modeling of the level lines $z = constant$ of integer values, that is, given n, to find those integer pairs (x, y) for which z is an integer also. As before, a computational problem of finding the largest and the smallest values for each of the variables x and y satisfying equation (16) gives rise to interesting activities on the use of inequalities.

Just like in the case of equation (5), the symmetric nature of equation (16) suggests that the inequalities $n < x \le y \le z$ need only be considered. Next, it might be helpful to start with generating integer values of z defined by equation (17) without regard to the computational efficiency of the environment. In that way, by modeling equation (17) for, say, $n = 2$ and $n =$

3, one can get some intuitive ideas as to what estimates for x and y might prove to be helpful in generating solutions for $n > 3$. In doing so, one can conjecture, that at most two denominators in the right-hand side of equation (16) may be greater than $3n$, and confirm this formally using proof by contradiction.

Proposition 7. Let $z \geq y \geq 3n$ in equation (16). Then the inequality $x \leq 3n$ holds true.

Proof. Assuming that, on the contrary, $x > 3n$ results in the false conclusion $\frac{1}{n}=\frac{1}{x}+\frac{1}{y}+\frac{1}{z}<\frac{1}{3n}+\frac{1}{3n}+\frac{1}{3n}=\frac{1}{n}$ thereby, establishing the following x-range

$$n+1 \leq x \leq 3n \tag{18}$$

Proposition 8. At most one of the variables x, y, z in equation (16) may be greater than $2n(n+1)$.

Proof. This time, we will use an *indirect* proof[8]—a combination of proof by contradiction and *constructive* proof where contradiction is constructed by calculating an appropriate example. To this end, one can choose $k>1$, $y = 2n(n+1)+1$, $z = 2n(n+1)+k$, and then construct a contradictory inequality for such values of k, y and z. Indeed,

$$\frac{1}{n}=\frac{1}{x}+\frac{1}{2n(n+1)+1}+\frac{1}{2n(n+1)+k}<\frac{1}{x}+\frac{2}{2n(n+1)+1}$$
$$=\frac{1}{x}+\frac{1}{n(n+1)+0.5}.$$

Therefore

$$\frac{1}{x}>\frac{1}{n}-\frac{1}{n(n+1)+0.5}=\frac{n^2+0.5}{n(n^2+n+0.5)}=\frac{1}{n(1+\frac{n}{n^2+0.5})}$$

whence $x<n+\frac{n^2}{n^2+0.5}<n+1$ as $\frac{n^2}{n^2+0.5}<1$. The conclusion $x<n+1$ contradicts to the inequality $x \geq n+1$, thus $y \leq 2n(n+1)$.

In order to find a lower estimate for y, spreadsheet calculations without regard to their efficiency can be helpful in conjecturing

Proposition 9. In equation (16) the inequality $y \geq 2n+1$ holds true.

Proof. Note that any representation of the fraction $\frac{1}{n}$ as a sum of three unit fractions can be generated by representing $\frac{1}{n}$ as a sum of two unit fractions, one of which, in turn, has to be represented as a sum of two unit fractions as well. In order to find a solution to equation

8 A classic demonstration of an indirect proof can be found in Pólya (1973).

(16) with the smallest possible value of y, one has to represent $\frac{1}{n}$ as the sum of two equal unit fractions,

$$\frac{1}{n} = \frac{1}{2n} + \frac{1}{2n},$$

and then represent $\frac{1}{2n}$ as the sum of the largest and the smallest possible unit fractions,

$$\frac{1}{2n} = \frac{1}{2n+1} + \frac{1}{2n(2n+1)}.$$

In that way, the solution of equation (16) in the form of the identity

$$\frac{1}{n} = \frac{1}{2n} + \frac{1}{2n+1} + \frac{1}{2n(2n+1)}$$

provides the smallest possible value of $y = 2n + 1$.

Therefore, the inequalities

$$2n + 1 \leq y \leq 2n(n + 1) \tag{19}$$

determine the y-range in equation (16).

Finally, one can formulate and prove

Proposition 10. In equation (16), none of the variables x, y, z may be greater than $n(n + 1)(n^2 + n + 1)$.

Proof. Applying identity (6) to itself yields

$$\frac{1}{n} = \frac{1}{n+1} + \frac{1}{n(n+1)} = \frac{1}{n+1} + \frac{1}{n(n+1)+1} + \frac{1}{n(n+1)[n(n+1)+1]}$$
$$= \frac{1}{n+1} + \frac{1}{n^2+n+1} + \frac{1}{n(n+1)(n^2+n+1)}.$$

That is,

$$\frac{1}{n} = \frac{1}{n+1} + \frac{1}{n^2+n+1} + \frac{1}{n(n+1)(n^2+n+1)} \tag{20}$$

In identity (20), $\frac{1}{n+1}$ and $\frac{1}{n(n+1)(n^2+n+1)}$ are, respectively, the largest and the smallest possible fractions among all triples of unit fractions into which the fraction $\frac{1}{n}$ can be partitioned.

Therefore, the inequalities

$$3n \leq z \leq n(n+1)(n^2+n+1) \tag{21}$$

determine the z-range in equation (16). To summarize our findings, variables x, y, and z in equation (16) satisfy, respectively, inequalities (18), (19), and (21).

Remark 4. Due to inequalities (18), variable x in equation (16) varies over the set $\{n+i \mid i=1,2,\ldots,2n\}$. Therefore, following Pólya's (1973) strategy of using a simpler problem as a means of understanding a more difficult one, equation (16) can be reduced to the following $2n$ equations

$$\frac{i}{n(n+i)} = \frac{1}{y} + \frac{1}{z}, \quad i = 1, 2, \ldots, 2n \tag{22}$$

	A	B	C	D	E	F	G	H	I	J
1	← →			y\x	4	5	6	7	8	9
2	n=	3		7			42	21		
3	x_{max}=	9		8		120	24		12	
4	y_{max}=	24		9		45	18			9
5	z_{max}=	156		10		30	15			
6				11						
7				12		20	12			
8				13	156					
9				14	84					
10				15	60	15				
11				16	48					
12				17						
13				18	36					
14				19						
15				20	30					
16				21	28					
17				22						
18				23						
19				24	24					

Figure 3.9. At most one denominator in each partition of 1/3 is greater than 24.

Some of equations (22) can be simplified to the form of equation (5) studied earlier. Such a reduction enables one to solve equation (22) for each i which divides $n(n+i)$ and, in doing so, to estimate from below the number of solutions of equation (16). For example, when n = 3, only two values of i turn equation (22) into equation (5) yielding, through such an approach, 13 solutions of equation (16) out of 21 total (see Figure 3.9). However, when $n = 2$, the reduction approach gives the same number of solutions to equation (16) as one can find through its direct modeling by a spreadsheet. This discussion leads to the problem of representing a non-unit fraction as a sum of two unit fractions discussed in detail in (Abramovich, 2006).

6. DISCUSSION OF MODELING DATA AND ITS ALTERNATIVE INTERPRETATION

The spreadsheet pictured in Figure 3.9 is designed to model equation (16)in the case n = 3 using inequalities (18) and (19). In that case, the inequalities give the ranges $4 \leq x \leq 9$ (*E1:J1*) and $7 \leq y \leq 24$ (*D2:D19*), respectively. The spreadsheet generates all partitions of 1/3 into three unit fractions. For example, using the contents of the cells *B2, E1, D8,* and *E8* as denominators of unit fractions yields the partition 1/3=1/4+1/13+1/156 with the largest product of the right-hand side's denominators among all 21 partitions so generated. Another partition, 1/3=1/9+1/9+1/9, is associated with the smallest such product. The two partitions confirm inequalities (21) when $n = 3$, namely, $9 \leq z \leq 156$. One can see that inequalities (21) were not used in the construction of the spreadsheet, except for establishing the maximum value of z. In that way, spreadsheet modeling can be used to confirm theoretical finding through a computational experiment.

By changing the content of cell *B2* (n-value), representations of other unit fractions as the sum of three like fractions (dimensions of corresponding rectangular prisms) can be generated and triples with specific properties can be identified.

Note that under certain conditions, the variables in equation (16) can be given another geometric interpretation: If n is the radius of a circle inscribed in a triangle with the heights x, y, and z, then the quadruple (n, x, y, z) satisfies equation (16). Indeed, if a, b and c are the sides of such a triangle then its area

$$A = n(\frac{a+b+c}{2}) = \frac{ax}{2} = \frac{by}{2} = \frac{cz}{2} \qquad (23)$$

whence equation (16). The chain of equalities (23) yields the relations

$$\frac{1}{x} = \frac{a}{2A}, \frac{1}{y} = \frac{b}{2A}, \frac{1}{z} = \frac{c}{2A}, \frac{1}{n} = \frac{a+b+c}{2A}.$$

Therefore,

$$\frac{1}{x} + \frac{1}{y} + \frac{1}{z} = \frac{a+b+c}{2A} = \frac{1}{n}.$$

However, not every integer solution of equation (16) corresponds to a triangle in which a circle of radius n is inscribed. In this regard, recall that the classic triangle inequality states that the sum of any two side lengths of a triangle is greater than the third side length. It is in that sense one may consider Heron's formula[9] for area of triangle

$$A = \sqrt{p(p-a)(p-b)(p-c)} \tag{24}$$

where $p = \dfrac{a+b+c}{2}$ as being defined for "all" a, b, and c. Using equalities (23) yield the following modification of Heron's formula

$$A = \left[\sqrt{\frac{1}{n}(\frac{1}{n} - \frac{2}{x})(\frac{1}{n} - \frac{2}{y})(\frac{1}{n} - \frac{2}{z})} \right]^{-1} \tag{25}$$

which right-hand side is defined for

$$z \geq y \geq x > 2n \tag{26}$$

Indeed, substituting $p = \dfrac{A}{n}, a = \dfrac{2A}{x}, b = \dfrac{2A}{y}, c = \dfrac{2A}{z}$ into formula (24) yields

$$A = \sqrt{\frac{A}{n}(\frac{A}{n} - \frac{2A}{x})(\frac{A}{n} - \frac{2A}{y})(\frac{A}{n} - \frac{2A}{z})} = A^2 \sqrt{\frac{1}{n}(\frac{1}{n} - \frac{2}{x})(\frac{1}{n} - \frac{2}{y})(\frac{1}{n} - \frac{2}{z})}$$

whence formula (25).

Therefore, one can see that only three out of 21 triples generated by the spreadsheet of Figure 3.9 satisfy inequalities (26); namely (7, 7, 21), (8, 8, 12) and (9, 9, 9). In the next section both plane and solid geometry contexts will be extended to allow for the use of arithmetic mean-geometric mean inequality in solving three-dimensional problems on maximum and minimum. In particular, through a combination of formal (the use of inequalities) and informal (the use of computing) arguments, it will be shown that the triples of heights (9, 9, 9) and (7, 7, 21) correspond, respectively, to triangles with the smallest and the largest area being circumscribed about a circle of radius three linear units.

[9] Heron (or Hero) of Alexandria—a Greek mathematician, physicist and inventor who lived probably in the 1st century A.D.

7. Formal and Informal Approaches to Solving Three-Dimensional Problems

One can use inequalities in solving problems on minimum and maximum in the context of three-dimensional modeling. As Kline (1985) put it: "A farmer who seeks the rectangle of maximum area with given perimeter might, after finding the answer to his question, turns to gardening, but a mathematician who obtains such a neat result would not stop there" (p. 133). In this regard, recall that one of the core recommendations of the Conference Board of the Mathematical Sciences (2001) regarding the mathematical education of teachers includes the need for courses that help teachers "develop the habits of mind of a mathematical thinker" (p. 8). Towards this end, a natural (from a mathematical point of view) extension of the activities described in the last section (as well as in the above quote about farming vs. doing mathematics) is to find rectangular prisms with the smallest and the largest surface area in the context of Problem 4. In order to find prism with the smallest surface area, one can utilize the three-dimensional case of the Arithmetic Mean—Geometric Mean inequality discussed in section 4.

Proposition 11. Let $u \geq 0,\ v \geq 0,\ w \geq 0$. Then

$$\frac{u+v+w}{3} \geq \sqrt[3]{uvw} \tag{27}$$

Proof. Inequality (27) shows that just like in the case of two non-negative numbers, the arithmetic mean of three non-negative numbers is greater than or equal to their geometric mean. A direct proof of inequality (27) may consist in reducing it to the form $\frac{p^3+q^3+r^3}{3} \geq pqr$, where $p=\sqrt[3]{u},\ q=\sqrt[3]{v},\ r=\sqrt[3]{w}$, and demonstrating that the difference between the left and right-hand sides of the last inequality can be expressed as a sum in which every term is obviously non-negative. To this end, using the formula $(a+b)^3 = a^3+b^3+3a^2b+3ab^2$ one can proceed as follows

$$\begin{aligned}(p+q+r)^3 &= (p+q)^3+r^3+3(p+q)^2r+3(p+q)r^2\\ &= p^3+q^3+r^3+3p^2q+3pq^2+3p^2r+3q^2r+6pqr+3pr^2+3qr^2\\ &= p^3+q^3+r^3+3pq(p+q+r)+3pr(p+q+r)+3qr(p+q+r)-3pqr.\end{aligned}$$

Therefore,

$$\begin{aligned}&p^3+q^3+r^3-3pqr\\ &=(p+q+r)^3-3pq(p+q+r)-3pr(p+q+r)-3qr(p+q+r)\\ &=(p+q+r)[(p+q+r)^2-3pq-3pr-3qr]\\ &=(p+q+r)(p^2+q^2+r^2-pq-pr-qr)\\ &=0.5(p+q+r)(p^2+q^2-2pq+p^2+r^2-2pr+q^2+r^2-2qr)\\ &=0.5(p+q+r)[(p-q)^2+(p-r)^2+(q-r)^2]\geq 0.\end{aligned}$$

In other words, $p^3+q^3+r^3 \geq 3pqr$ or $\frac{p^3+q^3+r^3}{3} \geq pqr$ whence $\frac{u+v+w}{3} \geq \sqrt[3]{uvw}$.

This completes the proof.

Remark 5. Alternatively, one can use *Maple* to demonstrate the correctness of inequality (27). Figure 3.10 shows a simple *Maple* code that carries out the above proof using symbolic computations.

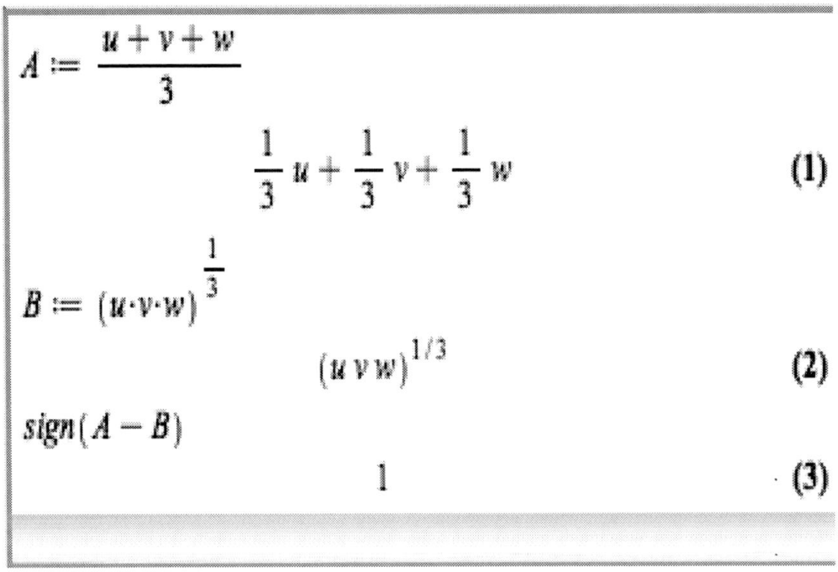

Figure 3.10. Verifying inequality (27) by using *Maple*.

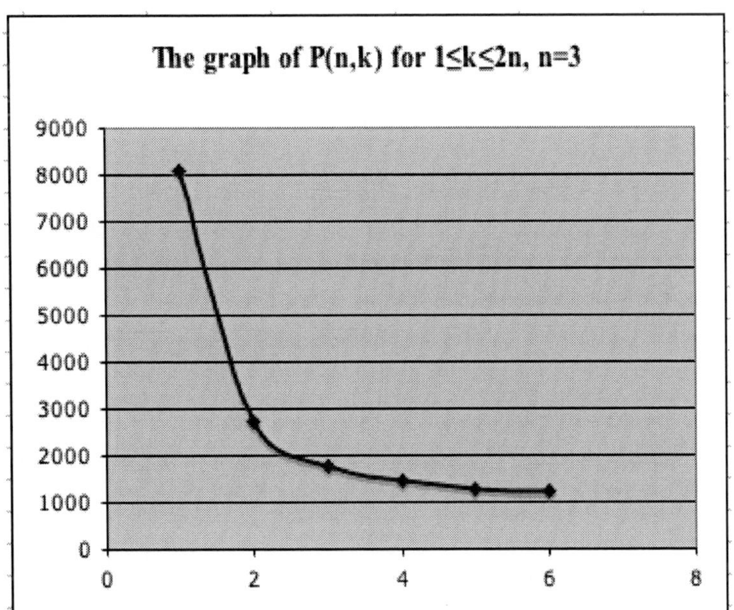

Figure 3.11. Visualizing the equality max $P(n,k)=P(n,1)$ for $n=3$.

Obviously, using *Maple* to assist in formal demonstration requires the development of new skills that, typically, are far from the basic ones. It appears, however, that the advent of computer algebra systems in the modern educational environment would not cause skills in rigorous mathematical proof to die out (as, for example, the skill of extracting square root taught in the schools until recently) but rather, proof assistant technology would be used by mathematicians and others as a means "to put the correctness of their proofs beyond reasonable doubt" (Harrison, 2008, p. 1405).

7.1. Finding the Prism with the Smallest Surface Area

Note that the surface area of the rectangular prism introduced through Problem 4 satisfies the equality $2(xy+yz+xz)=\frac{2xyz}{n}$. Therefore, those values of the variables x, y and z that provide the smallest product xyz (the volume of rectangular prism), bring about the smallest value for the surface area of the prism also. Substituting $x=\frac{1}{u}$, $y=\frac{1}{v}$, and $z=\frac{1}{w}$ in (27) results in the inequality $\frac{\frac{1}{x}+\frac{1}{y}+\frac{1}{z}}{3}\geq\sqrt[3]{\frac{1}{xyz}}$ whence $xyz\geq(\frac{3}{\frac{1}{x}+\frac{1}{y}+\frac{1}{z}})^3=27n^3$ with equality taking place when $x = y = z = 3n$. Thus, a cube with side $3n$ has the smallest surface area among all rectangular prisms whose volume is n times as much as half of this surface area. In particular, when $n = 3$ the triple (9, 9, 9) represents such a cube.

7.2. Finding the Prism with the Largest Surface Area Using a Combination of Formal Argument and Numeric Computation

Inequality (27) does not allow one to find the largest surface area among rectangular prisms satisfying the condition of Problem 4. In order to find such surface area, or, alternatively, the largest volume among the prisms, one can first find it numerically by using a spreadsheet. In doing so, one can discover that the denominators in the right-hand side of identity (20) provide the largest value for the product xyz; that is, these denominators—$n + 1$, $n^2 + n + 1$, $n(n + 1)(n^2 + n + 1)$—are the dimensions of the corresponding rectangular prism. Mathematically, this fact can be established through demonstrating that the function $V(n,k)=(n+k)^2[\frac{n}{k}(n+k)+1]^2\frac{n}{k}$, where $1\leq k\leq 2n$ is chosen to satisfy the identity

$$\frac{1}{n}=\frac{1}{n+k}+\frac{1}{\frac{n}{k}(n+k)+1}+\frac{1}{\frac{n}{k}(n+k)[\frac{n}{k}(n+k)+1]} \qquad (28)$$

(a generalization of identity [20]), monotonically decreases so that

$$V(n,k) \le V(n,1) = n(n+1)^2[n(n+1)+1]^2 \text{ for all } 1 \le k \le 2n.$$

In particular, when $n = 3$ the triple (4, 13, 156) represents dimensions of a rectangular prism with the largest surface area satisfying the condition of Problem 4. The graph of the function $V(n,k)$ for $n = 3$ is shown in Figure 3.11. It provides a visual demonstration of the monotonic descent of the function $V(3, k)$ for $1 \le k \le 6$.

7.3. A Three-Dimensional Geometric Problem in the Plane

A similar problem on minimum and maximum can be posed and solved in the context of the above-mentioned plane geometry interpretation of equation (16). To this end, among all triangles of integer heights circumscribed around a circle of radius n, one triangle has the smallest area and one triangle has the largest area. In order, to find the triangle with the smallest area one can set $u = p-a, v = p-b, w = p-c$, and apply inequality (27) to the right-hand side of formula (24) as follows

$$A \le \sqrt{p\left(\frac{p-a+p-b+p-c}{3}\right)^3} = \sqrt{\frac{p^4}{27}} = \frac{p^2}{3\sqrt{3}} \qquad (29)$$

Inequality (29) becomes an equality when $a = b = c$, the case of an equilateral triangle circumscribed about a circle with radius n. Equalities (23) imply that $A = pn$; thus, substituting A/n for p in the far-right part of (29) yields $A \le \frac{A^2}{3\sqrt{3}n^2}$ whence $A \ge 3\sqrt{3}n^2$ with equality taking place for an equilateral triangle with height and side equal, respectively, to $3n$ and $2\sqrt{3}n$. In particular, the case $n = 3$ yields $(x, y, z) = (9, 9, 9)$—one of the two triples mentioned at the end of section 6.

To find a circumscribed triangle with the maximum area, one can use a spreadsheet. Indeed, through spreadsheet modeling one can conjecture that the area of triangle in question attains the greatest value when $x = 2n + 1$, $y = 2n + 1$, $z = n(2n + 1)$. Using formula (25) for these values of x, y and z results in the formula

$$A = n^2(2n+1)\sqrt{\frac{2n+1}{2n-1}} \qquad (30)$$

In particular, when $n = 3$ one gets $(x, y, z) = (7, 7, 21)$—another triple mentioned at the end of section 6; one can find the area of a triangle with this triple of heights using the Pythagorean theorem. A formal demonstration of formula (30) requires the use of multivariable differential calculus that is beyond the scope of this book.

8. ACTIVITY SET

1. In the context of equation (3), prove the inequality $y \le n$ by contradiction using algebraic and contextual arguments.
2. In the context of equation (3), prove the inequality $y \le \frac{n}{b}$ by contradiction using algebraic and contextual arguments.
3. In the context of equation (3), using the assumptions $x \ge k$, $0 < k < n$, and the fact that y is an integer variable prove that $y \le INT(\frac{n-ak}{b})$.
4. Prove that none of the variables in equation (16) may be smaller or equal to n.
5. In the context of equation (16), prove that when $x = 2n + 1$, $y = 2n + 1$, and $z = n(2n + 1)$, formula (25) turns into formula (30).
6. Prove identity (28) using two approaches discussed in the context of proving identity (6).

Chapter 4

GEOMETRIC PROBABILITY

When it is not in our power to determine what is true, we ought to act according to what is most probable.

— Rene Descartes (1965, p. 21)

1. INTRODUCTION

This chapter reflects the author's experience in teaching probability topics to teachers in a capstone course following the belief that "to teach current school curricula, future high school teachers need to know more and somewhat different mathematics than mathematics departments have previously provided to teachers" (Conference Board of the Mathematical Sciences, 2001, p. 122). It reviews different technology-rich investigations on geometric probabilities emphasizing the role of mathematical connections, the use of multiple solution strategies, and the development of advanced mathematical thinking. Probability, a curriculum strand that starts in early grades and continues through high school and college, is closely connected to other strands dealing with number, geometry, and measurement (National Council of Teachers of Mathematics, 2000). In teaching probability concepts, the development of skill to assign number as a measure of geometrically expressed likelihood is an important didactical task. A geometric approach to probability requires the construction of regions in a sample space to serve as a model that represents the occurrence of a particular event followed by the exploration of numeric characteristics of those regions using mathematical tools that vary in complexity from proper fractions to definite integrals. One of the benchmarks for literacy in the STEM disciplines conceptualizes mathematical model as a tool that enables the demonstration and observation of different phenomena that go beyond one's intuition. The exploration of non-intuitive situations lends itself to the appropriate technological enhancement through the use of a spreadsheet, *The Geometer's Sketchpad* (*GSP*), *Maple*, and *Graphing Calculator 3.5* (*GC*).

Historically, the notion of geometric probability stems from the work of Buffon[1] on the development of fair games by introducing calculus into probability (Solomon, 1978). At a more basic level, the application of arithmetic to geometry brings about a numerical

[1] Georges-Louis Leclerc, de Buffon—a French naturalist, mathematician, and cosmologist of the 18th century.

representation of the likelihood of a particular event. We use to say, "there are three chances out of four" to pick up an even number from the set {1, 2, 4, 6}. However, how can one *measure* the likelihood (or chances) for such an event to realize? Here, the geometrization of the situation can naturally come into play if one represents the above set as a square divided into four equal parts (Figure 4.1). Then the measure of the part of the square filled with even numbers represents the numerical value of the likelihood of hitting a cell that includes an even number when a dart is thrown into the square. We assume that the dart always hits the square; in numerical terms, we assume that the square has unit area. Consequently, each of the four cells that comprise the square has area one-fourth.

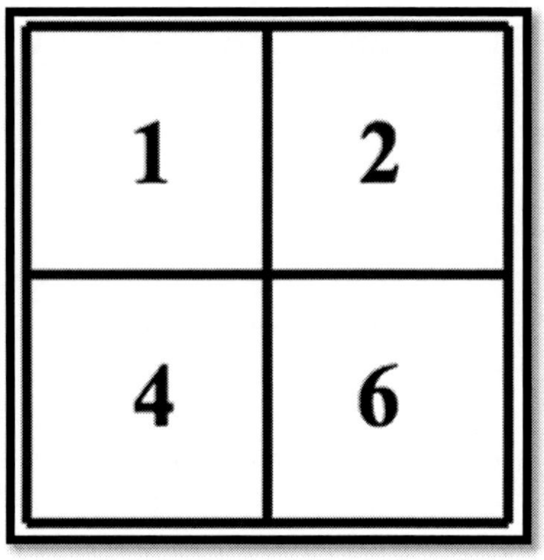

Figure 4.1. What fraction of the square is filled with even numbers?

Now, fractions can be used to measure chances if one defines the probability to hit a cell marked by an even number as fraction of the unit square marked by an even number. Clearly, the number 3/4 is such a fraction. In that way, the idea of geometric representation of fractions allows for their use as tools in finding the probability of a particular event (alternatively, the measure of the likelihood of the event).

2. Formal Definition of Geometric Probability

Consider a trial consisting in a random choice of a point over the region D. The task is to find the probability that the point belongs to the region d, which is a part of D. Let us assume that the outcomes of this trial are uniformly distributed over D. This assumption means that if the region D is decomposed into a finite number of equal sub-regions d_i and if E_i is the event that consists of falling a random point into the sub-region d_i then the outcomes E_i are equally likely occurrences. In other words, the probability $P(E_i)$ is proportional to the measure of d_i (length, area, volume) and does not depend on its position and shape.

Let E represent the following event: a point chosen randomly from D belongs to its part d. Then, by definition

$$P(E) = \text{measure}(d)/\text{measure}(D) \qquad (1)$$

In the example shown in Figure 4.1, E is the event that a point chosen randomly from the square of unit area belongs to its part d filled with even numbers. By the definition,

$$P(E) = \frac{3}{4} \div 1 = \frac{3}{4}.$$

3. Spreadsheet as a Geometric Medium

Many problematic situations modeled through the use of a spreadsheet can be extended into the probability curriculum strand enabling visualization capabilities of the software to be used in geometrizing numeric representations. As an example, consider

Exploration 1. A machine changed a 25-cent coin (a quarter) into dimes, nickels, and pennies. Assuming that it is equally likely to get any combination of the coins from the machine, find the probability that there are no pennies in the change.

	A	B	C	D
1	$n\backslash d$	0	1	2
2	0	**25**	**15**	**5**
3	1	**20**	**10**	**0**
4	2	**15**	**5**	
5	3	**10**	**0**	
6	4	**5**		
7	5	**0**		

Figure 4.2. What fraction of the (staircase-like) chart is shaded?

Reflections. The chart (generated by a spreadsheet) pictured in Figure 4.2 shows that there exist 12 ways to change a quarter using dimes, nickels, and pennies. For example, the triple of numbers located, respectively, in the cells *C1, A4, C4,* represents the following combination of coins: one dime, two nickels, and five pennies. Geometrically, region D filled with the number of pennies in each combination of coins consists of twelve regions d_i, $i = 1$,

2, …, 12. Let E be the event that a point randomly taken from region D belongs to one of the three regions d_i filled with a zero. Using formula (1) yields

$$P(E) = 3 \div 12 = \frac{1}{4}.$$

Remark 1. The spreadsheet representation of the context of Exploration 1 makes it possible to utilize conditional probabilities by considering vertical combinations of the cells (Figure 2) corresponding to the number of dimes in a change. Let E_i be the event that the change with no pennies includes i dimes, $i = 0, 1, 2$. Then $E = E_0 + E_1 + E_2$. We have

$$P(E_0) = \frac{1}{6} \cdot \frac{1}{2} = \frac{1}{12},\ P(E_1) = \frac{1}{4} \cdot \frac{1}{3} = \frac{1}{12},\ P(E_2) = \frac{1}{2} \cdot \frac{1}{6} = \frac{1}{12}.$$

Therefore,

$$P(E) = P(E_0) + P(E_1) + P(E_2) = \frac{3}{12} = \frac{1}{4}.$$

In a similar way, let $\hat{E}_i$ be the event that the change with no pennies includes i nickels, i = 0, 1, 2, 3, 4, 5. Note that $P(\hat{E}_i) = 0$ for i = 0, 2, 4. At the same time

$$P(\hat{E}_1) = \frac{1}{3} \cdot \frac{1}{4} = \frac{1}{12},\ P(\hat{E}_3) = \frac{1}{2} \cdot \frac{1}{6} = \frac{1}{12},\ P(\hat{E}_5) = 1 \cdot \frac{1}{12} = \frac{1}{12}.$$

Therefore, once again,

$$P(E) = P(\hat{E}_1) + P(\hat{E}_3) + P(\hat{E}_5) = \frac{3}{12} = \frac{1}{4}.$$

This alternative approach to Exploration 1 supports the recommendation that teachers need to "understand basic concepts of probability such as conditional probability and independence, and develop skill in calculating probabilities associated with those concepts" (Conference Board of the Mathematical Sciences, 2001, p. 44).

One can use a spreadsheet to construct different geometric designs (shapes) and then use those shapes in the context of dart games. Connecting geometric probabilities to dart games makes it possible to integrate probability strand with topics in algebra and discrete mathematics (including number theory). These topics may include polygonal numbers, summation techniques, and algebraic inequalities. With this in mind, consider

Exploration 2. Jeremy is building towers from blocks according to the pattern shown in Figure 4.3 and then throw a dart into a tower. Assuming that the dart would always hit the minimal rectangular enclosure within which the corresponding tower can be placed, what is the probability that the dart would miss the tower? What is the smallest tower for which the probability becomes smaller than 1/10?

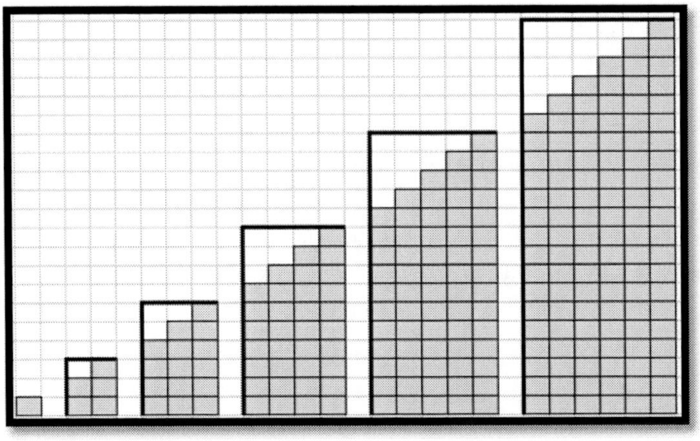

Figure 4.3. Jeremy's towers placed in rectangular enclosures.

Reflections. One can observe that the sequence

1, 2 + 3, 4 + 5 + 6, 7 + 8 + 9 + 10, 11 + 12 + 13 + 14 + 15,
16 + 17 + 18 + 19 + 20 + 21

describes the number of blocks used to build the six towers shown in Figure 4.3. Each sum, beginning from the second, starts with a number that is one greater than the corresponding triangular number. Indeed, 2 = 1 + 1, 4 = 3 + 1, 7 = 6 + 1, 11 = 10 + 1, and 16 = 15 + 1, where 1, 3, 6, 10, and 15 are consecutive triangular numbers. Let t_n denote the triangular number of rank n. Then the n-th tower, having the width n, consists of the following number of blocks

$$t_{n-1}+1+t_{n-1}+2+...+t_{n-1}+n=nt_{n-1}+t_n=\frac{n^2(n-1)}{2}+\frac{n(n+1)}{2}=\frac{n(n^2+1)}{2}.$$

Here the formula $t_n=\frac{n(n+1)}{2}$ (introduced in Chapter 1) was used. The minimal rectangular enclosure includes $t_n n=\frac{n^2(n+1)}{2}$ blocks. Therefore, according to definition (1), the probability of missing the n-th tower is equal to

$$P(n)=\frac{\frac{n^2(n+1)}{2}-\frac{n(n^2+1)}{2}}{\frac{n^2(n+1)}{2}}=\frac{n-1}{n(n+1)}.$$

One can see that $P(n+1)<P(n)$ for all $n > 2$ and $\lim_{n\to\infty}P(n)=0$. That is, the probability to miss a tower monotonically decreases and tends to zero as n becomes greater than two. Indeed,

$$P(n+1)-P(n)=\frac{n}{(n+1)(n+2)}-\frac{n-1}{n(n+1)}=\frac{n^2-(n+2)(n-1)}{n(n+1)(n+2)}$$
$$=\frac{n^2-(n^2+n-2)}{n(n+1)(n+2)}=-\frac{n-2}{n(n+1)(n+2)}<0.$$

That is, $P(n + 1) < P(n)$ for all $n > 2$ (note: $P(2) = P(3) = 1/6$) and

$$\lim_{n\to\infty} P(n)=\lim_{n\to\infty}\frac{n-1}{n(n+1)}=\lim_{n\to\infty}\frac{1-\frac{1}{n}}{n(1+\frac{1}{n})}=0.$$

In order to find the smallest n (that is, the smallest tower) for which the probability becomes smaller than 1/10, one has to solve the inequality

$$\frac{n-1}{n(n+1)}<\frac{1}{10} \tag{2}$$

Inequality (2), due to the obvious assumption $n > 0$, is equivalent to $10n-10<n^2+n$ or $n^2-9n+10>0$. The use of the quadratic formula yields the inequality $n>\frac{9+\sqrt{41}}{2}$. As $7<\frac{9+\sqrt{41}}{2}<8$, one can conclude that $n = 8$ is the smallest whole number for which inequality (2) holds true. In other words, the 8th tower is the smallest one for which the probability to miss it is smaller than 1/10.

Remark 2. In much the same way, other types of towers can be constructed[2] and different polygonal numbers can be used as problem-solving tools. For example, the arrangement of the elements of towers so that the widths of the towers are represented by the sequence 1, 3, 5, 7, 9, …, $2n - 1$, will lead to the joint use of triangular and square (s_n) numbers; when the widths are 1, 4, 7, 10, 13, ..., $3n - 2$—triangular and pentagonal (p_n) numbers, and so on. Indeed, in the first case, each of the sums (representing the number of blocks used to construct the first five towers)

1, 2 + 3 + 4, 5 + 6 + 7 + 8 + 9, 10 + 11 + 12 + 13 + 14 + 15 + 16,
17 + 18 + 19 + 20 + 21 + 23 + 24 + 25 + 26, 27 + …

starts with a number (2, 5, 10, 17, …) that is one greater than a square number; in the second case, each of the sums (representing the number of blocks used to construct the first five towers)

[2] One can program a spreadsheet to automatically generate different types of towers controlled by the properties of polygonal numbers. The corresponding programming details are rather advanced and can be found elsewhere (Abramovich, 2003; Abramovich and Sudgen, 2004).

$$1, 2 + 3 + 4 + 5, 6 + 7 + 8 + 9 + 10 + 11 + 12,$$
$$13 + 14 + 15 + 16 + 17 + 18 + 19 + 20 + 21 + 22, 23 + \ldots$$

starts with a number (2, 6, 13, 23, ...) that is one greater than a pentagonal number. Then the n-th tower in the first case consists of $ns_{n-1} + t_n$ blocks; the n-th tower in the second case consists of $np_{n-1} + t_n$ blocks. In that way, new problems on geometric probability can be explored.

x o	x o	x o	x o	x	x	x
x o	x o	x o	x o	x	x	x
x o	x o	x o	x o	x	x	x
o	o	o	o			
o	o	o	o			

Figure 4.4. From more x's than o's to same number of the letters in the non-shaded area.

4. Comparing Probabilities of Two Events through Geometrization

The idea of geometrization of probabilities makes it possible not only to give meaning to problem solving associated with finding the measures of a likelihood but in addition, using a grade-appropriate context, to discover hidden properties of fractions. With this in mind, consider

The M&Ms Brainteaser 1. Billy wants to eat red M&Ms only. There are two bags of M&Ms available, plain and peanut. If there are 3 red plain out of 5 total and 4 red peanut out of 7 total, for which bag does Billy have higher probability to get a red M&M?

Reflections. Arithmetically, the question to be answered is: Which of the two fractions, 3/5 or 4/7, is bigger? Such a comparison can be carried out geometrically as shown in Figure 4.4. The fraction of the grid filled with x's is equal to 3/5; the fraction of the grid filled with o's is equal to 4/7. Counting x's yields 21; counting o's yields 20. Thus, the inequality $21 > 20$ implies $3/5 > 4/7$. Contextually, the last inequality means that there is a higher probability to pick up a red plain M&M than a red peanut M&M.

Remark 3. The grid pictured in Figure 4.4 has an interesting property. If one removes from the grid its shaded part consisting of the far-left column and the top row, the remaining number of x's and o's would be the same. Furthermore, if one applies the same operation to the new (non-shaded) grid, the number of x's would become smaller than the number of o's. Such a property of the grid can be utilized in posing new geometric probability questions. For example, one can extend the above brainteaser to involve more than one person dealing with the bags of M&Ms.

The M&Ms Brainteaser 2. Billy and Mary want to eat red M&Ms only. There are two bags of M&Ms available, plain and peanut. If there are 3 red plain out of 5 total and 4 red peanut out of 7 total, how many red M&Ms of each kind does Billy have to eat in order to allow Mary enjoying higher probability to get a red peanut M&M that a red plain one?

Reflections. Arithmetically, the following three fractional relationships provide a resolution to the second brainteaser: 3/5 > 4/7, 2/4 = 3/6, and 1/3 < 2/5. Contextually, Billy has to eat two red M&Ms from each bag in order to give Mary more chances to pick up a red peanut M&M than a red plain M&M so that the first bag has one red M&M out of three total and the second bag has two red M&Ms out of five total. Geometrically, one can see that the situation shown in Figure 4.5 where the grid has more *x*'s (twenty-one) than *o*'s (twenty), but after removing two differently shaded parts of the grid, it remains (in the non-shaded area) more *o*'s (six) than *x*'s (five).

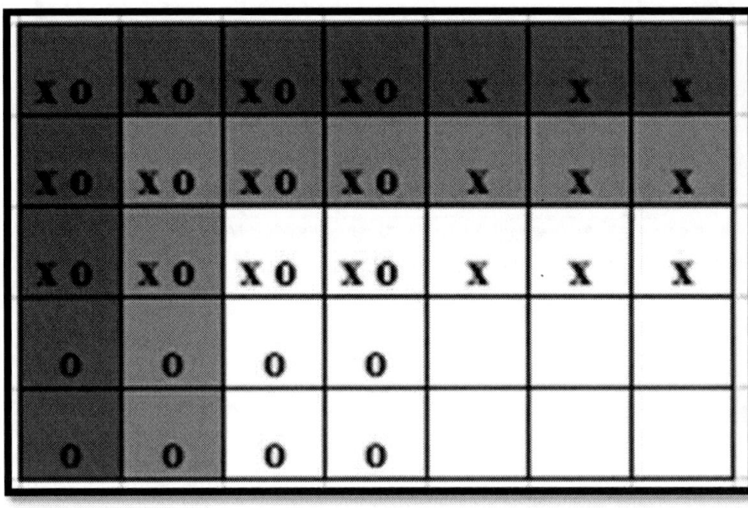

Figure 4.5.The second elimination of the flipped *L* shape results in fewer *x*'s than *o*'s.

Remark 4. Interestingly, the geometric situation presented in Figure 4.5 is not easy to replicate without some kind of generalization. In algebraic terms, one has to find two proper non-unit fractions $\frac{a}{b}$ and $\frac{c}{d}$ so that

$$\frac{a}{b} > \frac{c}{d} \text{ and } \frac{a-1}{b-1} = \frac{c-1}{d-1}.$$

A simple observation shows that in both fractions, 3/5 and 4/7, denominator is one less than twice the corresponding numerator. In order to generalize from this observation, let $1 < a < b$. One can check to see that

$$\frac{a}{2a-1} > \frac{b}{2b-1} \text{ and } \frac{a-1}{2a-2} = \frac{b-1}{2b-2} = \frac{1}{2}.$$

Now, other pairs of fractions can be generated such as

5/9 and 6/11 for which 5/9 > 6/11, 4/8 = 5/10, 3/7 < 4/9;

7/13 and 9/17 for which 7/13 > 9/17, 6/12 = 8/16, 5/11 < 7/15;

and so on. Furthermore, given a proper fraction $\frac{a}{b}$, one can generate an infinite number of fractions satisfying the situation described by the second brainteaser.

Proposition 1. Let $b > a > 2$ and $k > 1$. Then

$$\frac{a}{b} > \frac{k(a-1)+1}{k(b-1)+1}$$

and

$$\frac{a-2}{b-2} < \frac{k(a-1)-1}{k(b-1)-1}$$

Proof. As $b > a$ and $k > 1$, we have

$$\frac{a}{b} - \frac{k(a-1)+1}{k(b-1)+1} = \frac{ak(b-1)+a-bk(a-1)-b}{b(k(b-1)+1)} = \frac{(b-a)(k-1)}{b(k(b-1)+1)} > 0.$$

Next, for all $k > 1$, the relation

$$\frac{a-1}{b-1} = \frac{k(a-1)+1-1}{k(b-1)+1-1}$$

is true. Finally, as $b > a > 2$ and $k > 1$, we have

$$\frac{k(a-1)-1}{k(b-1)-1} - \frac{a-2}{b-2} = \frac{(ka-k-1)(a-2)-(kb-k-1)(b-2)}{(k(b-1)-1)(b-2)}$$
$$= \frac{2k(b-a)-(k+1)(b-a)}{(k(b-1)-1)(b-2)} = \frac{(b-2)(k-1)}{(k(b-1)-1)(b-2)} > 0.$$

This completes the proof of Proposition 4.1.

Example 1. When $a = 3$, $b = 5$, and $k = 3/2$ the pair 3/5 and 4/7 results from Proposition 1.

In that way, we have proved what may be referred to

The 1st M&M Probability Theorem (PT1). There are two bags of M&Ms, plain and peanut, respectively, with a red plain M&Ms out of b total and $k(a-1)+1$ red peanut M&Ms out of $k(b-1)+1$ total. Let $b > a > 2$ and $k > 1$. Then

1. The probability to pick up a red plain M&M is higher that the probability to pick up a red plain M&M.
2. After eating one red M&M from each bag, the probability to pick up a red plain M&M becomes equal to the probability to pick up a red peanut M&M.
3. One has to eat one more red M&M from each bag in order to have higher probability to pick up a red peanut M&M.

Remark 5. The following question remains to be answered: Does PT1 provide necessary and sufficient conditions for the existence of such bags of M&Ms or does the theorem provide just sufficient conditions? In other words, does the algorithm described by PT1 generate all pairs of fractions $\frac{a}{b}$ and $\frac{c}{d}$ for which $\frac{a}{b} > \frac{c}{d}$, $\frac{a-1}{b-1} = \frac{c-1}{d-1}$, and $\frac{a-2}{b-2} < \frac{c-2}{d-2}$, where $b > a > 2$ and $d > c > 2$?

Remark 6. Furthermore, one may wonder, is it is possible to find two proper non-unit fractions $\frac{a}{b}$ and $\frac{c}{d}$ such that $\frac{a}{b} > \frac{c}{d}$ yet $\frac{a-1}{b-1} < \frac{c-1}{d-1}$? (Below, such pairs of fractions will be referred to as "jumping" fractions). In other words, do there exist two bags of M&Ms so that the corresponding chances (probabilities) to pick up an M&M of the specified color can be reversed after eating one candy of that color from each bag? A trial and error approach (enhanced by spreadsheet computing) can yield the following quadruple: $(a, b, c, d) = (6, 32, 4, 20)$. Indeed, $\frac{4}{20} > \frac{6}{32}$ and $\frac{3}{19} < \frac{5}{31}$. In terms of the context, one can conclude that although originally, for the bags with 6 red plain M&Ms out of 32 total and 4 red peanut M&Ms out of 20 total, the chances are in favor of red peanut M&M, Billy can reverse the chances after eating just one red M&M from each bag.

Two new questions can be formulated in this regard.

Question 1. How can one find such pairs of bags in a systematic way rather than through trial and error?

Question 2. Do there exist pairs of bags for which the chances to pick up an M&M of the specified color could be reversed more than one time?

In the next two sections we will attempt to answer these questions.

5. A Systematic Approach to Generating "Jumping" Fractions

To answer the first question, consider two fractions $\frac{1}{b}$ and $\frac{c}{bc+k}$ where b, c, and k are positive integers. Using cross-multiplication, one can show that

$$\frac{1}{b} > \frac{c}{bc+k}$$

and, therefore, for any $a \neq 0$

$$\frac{a}{ab} > \frac{c}{bc+k} \tag{3}$$

Proposition 2. Let $a > 1, b > 1, c > a,\ k < \dfrac{(b-1)(c-a)}{a-1}$. Then

$$\frac{a-1}{ab-1} < \frac{c-1}{bc+k-1} \tag{4}$$

In other words, the fractions $\dfrac{a}{ab}$ and $\dfrac{c}{bc+k}$ are "jumping" fractions.

Proof.

$$\frac{c-1}{bc+k-1} - \frac{a-1}{ab-1} = \frac{(c-1)(ab-1)-(a-1)(bc+k-1)}{(bc+k-1)(ab-1)}$$
$$= \frac{abc-c-ab+1-abc-ak+a+bc+k-1}{(bc+k-1)(ab-1)}$$
$$= \frac{(c-a)(b-1)-k(a-1)}{(bc+k-1)(ab-1)} > \frac{(c-a)(b-1)-(c-a)(b-1)}{(bc+k-1)(ab-1)} = 0.$$

Example 2. When $a = 4$, $b = 5$, $c = 6$, and $k = 2 < \dfrac{(5-1)(6-4)}{(4-1)} = \dfrac{8}{3}$, the pair of "jumping" fractions $\dfrac{4}{20}$ and $\dfrac{6}{32}$ result from Proposition 5.1.

In that way, we have proved what may be referred to

The 2nd M&M Probability Theorem (PT2). There are two bags of M&Ms, plain and peanut, respectively, with a red plain M&Ms out of ab total and c red peanut M&Ms out of $bc + k$ total. Let $a > 1, b > 1, c > a,\ k < \dfrac{(b-1)(c-a)}{a-1}$. Then

1. The probability to pick up a red plain M&M is higher than the probability to pick up a red peanut M&M.
2. One has to eat one red M&M from each bag in order to have higher probability to pick up a red peanut M&M than a red plain M&M.

Remark 6. PT2 provides only sufficient conditions for the existence of two bags of M&Ms described by inequalities (3) and (4). These inequalities are not necessary for the existence of a pair of "jumping" fractions. In other words, inequalities (3) and (4) do not describe all possible grids where after a single operation of removing the top row and the far-left column geometric probabilities are reversed. For example,

$$\frac{4}{19} > \frac{6}{30} \text{ and } \frac{3}{18} < \frac{5}{29}.$$

Yet, the "jumping" fractions $\frac{4}{19}$ and $\frac{6}{30}$ do not follow the algorithm of PT2, thereby, demonstrating that whereas PT2 does give an algorithm for generating certain bags of M&Ms, other bags exist that do not satisfy PT2.

6. Explaining the Behavior of Fractions through the Behavior of Functions

Now, we will attempt to answer the second question posed at the end of section 4. As before, recourse to geometry can clarify the situation and, in addition, provide a kind of an algorithm for generating different bags not included into PT2. To this end note that the probability to pick up a red M&M for each bag can be associated with the rational linear functions $f(x)=\frac{a-x}{b-x}$ and $g(x)=\frac{c-x}{d-x}$ which for $a < b$ and $c < d$ decrease monotonically, that is, $f(x + 1) < f(x)$ as x grows large. Indeed,

$$f(x+1)-f(x)=\frac{a-x-1}{b-x-1}-\frac{a-x}{b-x}=\frac{a-b}{(b-x-1)(b-x)}<0.$$

Therefore, in order for the pair of fractions $\frac{a}{b}$ and $\frac{c}{d}$ be "jumping" fractions, the graphs of the functions $f(x)$ and $g(x)$ should intersect in the interval (0, 1). From the equation $\frac{a-x}{b-x}=\frac{c-x}{d-x}$ it follows that $x=\frac{ad-bc}{a+d-b-c}$ is the point of intersection of the two functions. This leads to the inequality

$$0<\frac{ad-bc}{a+d-b-c}<1 \tag{5}$$

from which the appropriate quadruples (a, b, c, d) of, generally speaking, rational numbers, can be chosen, although not without difficulty of dealing with four variables. Finally, because the graphs of two linear rational functions can intersect at one point only, the probabilities cannot be reversed more than one time; that is, the "jumping" effect can be observed only once.

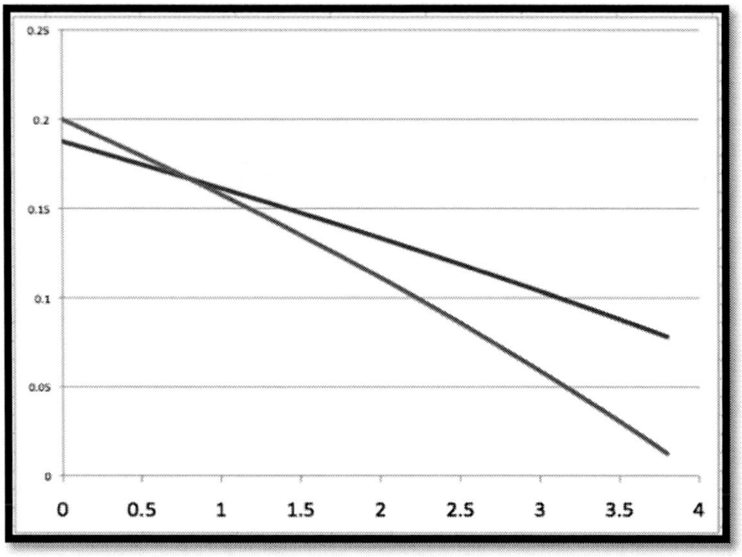

Figure 4.6. The graphs of $f(x)$ and $g(x)$ intersect in the interval (0, 1).

Figure 4.6 shows the graphs of the functions $f(x) = \frac{4-x}{20-x}$ and $g(x) = \frac{6-x}{32-x}$ where $f(0) = \frac{4}{20}$, $f(1) = \frac{3}{19}$, $g(0) = \frac{6}{32}$, and $g(1) = \frac{5}{31}$. That is, whereas $f(0) > g(0)$, the inequality $f(1) < g(1)$ takes place and cannot be reversed as x becomes greater than one. Note that the fractions $\frac{4}{20}$ and $\frac{6}{32}$ satisfy the conditions of PT2. On the other hand, the fractions $\frac{4}{19}$ and $\frac{6}{30}$ do not satisfy conditions of PT2; yet, the functional perspective brings about the functions $f_1(x) = \frac{4-x}{19-x}$ and $g_1(x) = \frac{6-x}{30-x}$ for which $f_1(0) > g_1(0)$ and $f_1(1) < g_1(1)$. Alternatively, inequalities (5) turn into $0 < \frac{4 \cdot 30 - 19 \cdot 6}{4 + 30 - 19 - 6} < 1$ or $0 < \frac{2}{3} < 1$, which is true. So, whereas inequalities (5) provide necessary and sufficient conditions for the existence of a pair of "jumping" fractions, inequalities (3) and (4) are only sufficient but not necessary conditions for the existence of such fractions.

Another approach to generating "jumping" fractions is to replace two parameters in inequalities (5) by numbers and then, using the *GC*, graph regions in the plane of two other parameters where inequalities (5) are satisfied. Finally, one has to select either integer or rational points that belong to these regions. For example, when $a = 6$ and $c = 4$, inequalities (5) turn into

$$0 < \frac{6d - 4b}{2 + d - b} < 1 \qquad (6)$$

Setting $b = x$, $d = y$, one can graph the inequality $\sqrt{\frac{6y-4x}{2+y-x}}\sqrt{1-\frac{6y-4x}{2+y-x}} > 0$ as the locus of inequalities (6) in the plane (x, y) of the *GC*.

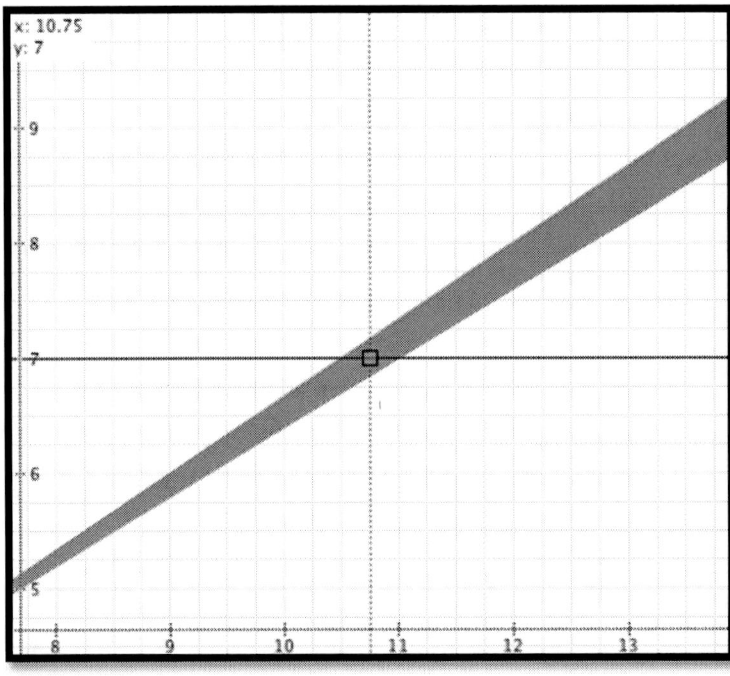

Figure 4.7. Constructing the locus of inequalities (6).

The shaded region shown in Figure 4.7 allows one to make several observations. First, the region is located between two straight lines the slopes of which differ by 1/15, thereby, it cannot be characterized by the abundance of points with integer coordinates. Second, by selecting $d = 7$, one can see that the closest integer values of b, 10 and 11, intersect the line $d = 7$ outside the region. Third, one can select $b = 10.75$ (among infinitely many other rational values of b located between 10 and 11) which, therefore, along with $d = 7$ brings about two fractions $\frac{4}{7}$ and $\frac{24}{43}$ so that $\frac{4}{7} > \frac{24}{43}$ yet $\frac{3}{6} < \frac{23}{42}$.

In much the same way, other "jumping" fractions already mentioned above can be generated. For example, by setting $d = 20$ one can find two integer points (32, 20) and (31, 20) that belong to the region shown in Figure 4.7. This gives two pairs of "jumping" fractions, $(\frac{4}{20}, \frac{6}{32})$ and $(\frac{4}{20}, \frac{6}{31})$.

7. Functions with Parameters and Geometric Probability

Using the analysis of functions to explain numeric phenomena related to geometric probability can motivate the introduction of new exploratory activities associated with evaluating the probability of a certain behavior of functions using a geometric approach. For example, it is unlikely to find bags of M&Ms with the distribution of candies stated in either PT1 or PT2 unless a systematic approach is developed. As noted by the Committee on K-12 Engineering Education (Katehi, Pearson, and Feder, 2009), "systems may have unexpected effects that cannot be predicted from the behavior of individual subsystems." Indeed, in designing systems depending on parameters it is important to know the likelihood of a certain phenomenon to occur if parameters are to be chosen at random from a certain set. With this in mind, consider

Exploration 3. There are two bags of M&Ms, plain and peanut, respectively, with 6 red plain M&Ms out of a total and 4 red peanut M&Ms out of b total. If a pair (a, b) is chosen at random from the rectangle

$$6 < a < R, 4 < b < R \tag{7}$$

what is the probability of having smaller chances to pick up a red plain M&M than a red peanut M&M after one red candy from each bag is eaten?

Reflections. Consider the functions $f(x) = \frac{4-x}{b-x}$ and $g(x) = \frac{6-x}{a-x}$ for which the inequalities $f(0) > g(0)$ and $f(1) < g(1)$ hold. In terms of parameters a and b, the last two inequalities can be written as $\frac{4}{b} > \frac{6}{a}$ and $\frac{3}{b-1} < \frac{5}{a-1}$. This leads to the system of linear inequalities

$$b < \frac{2}{3}a\,,\ b > \frac{3}{5}a + \frac{2}{5} \tag{8}$$

The regions defined by inequalities (7) and (8) are shown in Figure 4.8 for $R = 12$.

In order to find the area of the dark region that belongs to the rectange from which the pair (a, b) is chosen (Figure 4.8), one has to compute

$$\int_6^R \left(\frac{2}{3}a - \frac{3}{5}a - \frac{2}{5}\right)da = \frac{1}{15}\int_6^R (a-6)da = \frac{(a-6)^2}{30} = \frac{(R-6)^2}{30}.$$

If point (a, b) belongs to the dark region, then the desired event occurs. As the area of the rectangle defined by inequalities (7) is equal to $(R - 6)(R - 4)$, the probability $P(R)$ of this event can be found by using formula (1)

$$P(R) = \frac{(R-6)^2}{30(R-6)(R-4)} = \frac{R-6}{30(R-4)}.$$

In particular, when $R = 12$ we have $P(12) = 0.025$. Such a low probablity explains as to why it was difficult to find a pair of fractions satisfying conditions of the second brainteaser.

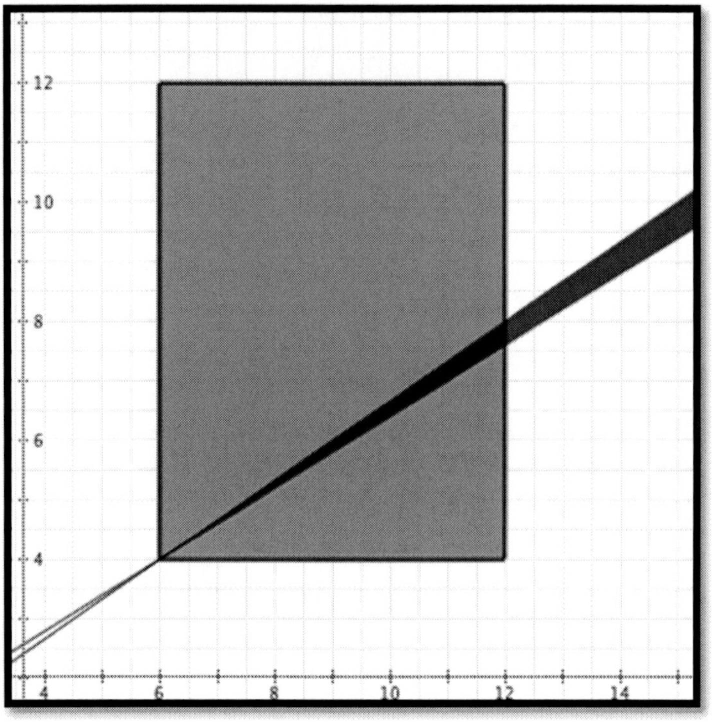

Figure 4.8. What is the likelihood to choose a point (a, b) from the dark region?

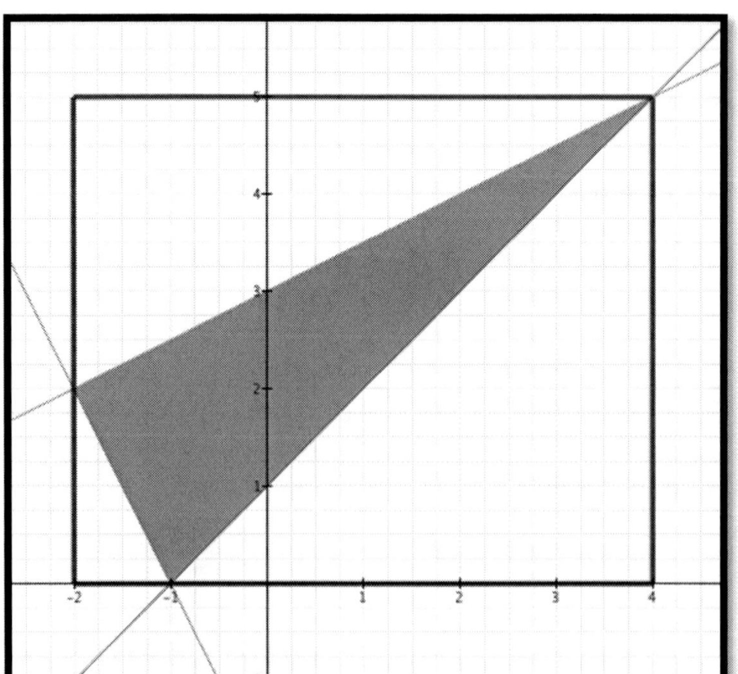

Figure 4.9. Triangular region of parameters responsible for the behavior of $f(x)$.

Exploration 4. Parameters a and b are chosen at random over the rectangle with the vertices (-2, 0), (-2, 5), (4, 5), (4, 0). Find the probability that the linear function $f(x) = ax + b$ at the points $x = 2$, $x = -1$, and $x = -0.5$ assumes values such that

$$f(2) > -2, f(-1) > 1, f(-0.5) < 3 \tag{9}$$

Reflections. In order to be able to apply definition (1) to calculating the probability of the behavior of the function $f(x)$ with parameters a and b randomly chosen from the given rectangle, one has to formulate this behavior in terms of the region from which these parameters (slope-intercept characteristics) are selected. Then the area of the part of the region that belongs to the rectangle when divided by the area of the rectangle, A_{REC}, would yield the probability sought. One approach is to use the *GC*. To this end, note that inequalities (9) are equivalent, respectively, to the following three inequalities in the plane of parameters a and b

$$2a + b > -2, -a + b > 1, -0.5a + b < 3 \tag{10}$$

Replacing inequalities (10) by equalities

$$b = -2a - 2, b = 1 + a, b = 0.5a + 3 \tag{11}$$

in which parameter b is expressed as a linear function of parameter a and constructing in the coordinate plane (a, b) the graphs of equations (11) along with the rectangle defined by its vertices, results in the triangle all points of which belong to the rectangle (Figure 4.9).

There are several ways in which the area of the triangle, $A_{TR.}$, can be found. The first way is to use traditional integration skills:

$$A_{TR.} = \int_{-2}^{-4}(0.5a + 3)da - \int_{-2}^{-1}(-2a - 2)da - \int_{-1}^{4}(a + 1)da = \frac{15}{2}.$$

The second way is to outsource the integration to mathematical software, like *Maple*, as shown in Figure 4.10.

The third way is to replicate Figure 4.9 in the setting of the *GSP* and then calculate area using measuring capabilities of the program. Note that all three approaches give the same result, $A_{TR} = 7.5$. As the area of the rectangle from which parameters a and b are chosen is equal to 30, the probability sought can be found as the ratio

$$\frac{A_{TR.}}{A_{REC.}} = \frac{7.5}{30} = \frac{1}{4}.$$

This completes the geometric (theoretical) approach to Exploration 4. In the next section, it will be shown how this theoretical approach can be complemented by an experimental approach to calculating the required probability.

$$Area := \int_{-2}^{4} (0.5 \cdot a + 3)\, da - \int_{-2}^{-1} (-2 \cdot a - 2)\, da - \int_{-1}^{4} (1 + a)\, da$$

$$7.50000000$$

Figure 4.10. *Maple* integration in the case of Exploration 4.

8. CALCULATING GEOMETRIC PROBABILITY THROUGH A SPREADSHEET-BASED SIMULATION

A conceptually rich extension of Exploration 4 is to compare theoretical and experimental probabilities of the event formulated in terms of the behavior of the function $f(x) = ax + b$ with random parameters a and b. An experimental probability of an event (alternatively, the frequency of a random event) is defined as the ratio of the number of times the event occurred in a series of identical trials to the total number of trials in this series. As the number of trials becomes sufficiently large, the difference between the two probabilities typically becomes insignificant. In other words, the frequency of the occurrence of a random event approaches the probability of this event as the number of trials increases.

	A	B	C	D	E	F	G	H	I	J	K
1	a	b		f(2)+2	f(-1)-1	3-f(-0.5)		Exp. Probability			
2	3.62	4.34		13.5836329	-0.28	0.47		235	0.24	1	0.252
3	3.13	2.74		11.0010087	-1.40	1.83				2	0.262
4	2.69	2.71		10.0804161	-0.98	1.64				3	0.234
5	-1.05	0.62		0.52464922	0.67	1.85	1			4	0.234
6	0.40	2.07		4.85747646	0.67	1.13	1			5	0.27
7	1.06	0.98		5.1015989	-1.08	2.55				6	0.251
8	-1.62	3.68		2.45338466	4.30	-1.49				7	0.266
9	-1.76	2.54		1.02990554	3.30	-0.42				8	0.24
10	2.27	2.03		8.57340649	-1.24	2.11				9	0.26
11	0.04	2.36		4.43841807	1.32	0.66	1			10	0.248
12	1.24	1.96		6.44336848	-0.28	1.66				11	0.235
13	-1.17	2.84		2.5076589	3.01	-0.42				12	0.235
14	-0.42	1.75		2.91043105	1.17	1.04	1			13	0.245
15	2.11	2.26		8.47535114	-0.84	1.79				14	0.233
16	2.12	3.12		9.35334312	0.00	0.94	1			15	0.249
17	-0.07	0.75		2.60341462	-0.18	2.22				16	0.272
18	1.48	0.86		5.81409839	-1.62	2.88				17	0.253
19	0.07	0.99		3.13331013	-0.07	2.04				18	0.276
20	-1.93	1.75		-0.11958809	2.68	0.28				19	0.246
1001	2.02	1.87		7.89634293	-1.15	2.14					

Figure 4.11. Calculating experimental probability for Exploration 4.

To demonstrate the manifestation of this mathematically deep phenomenon (in more general terms known as the law of large numbers), one can use a spreadsheet. To this end, using the spreadsheet function = (Y – X)*RAND() + X that randomly generates a number from the interval (X, Y), one can select, say, 10^3 points that belong to the region $\{(a,b) \mid -2 < a < 4, 0 < b < 5\}$ (Figure 4.11, columns *A* and *B*). Then for each such pair of parameters (a, b) calculate the values of $f(2) + 2$, $f(-1) - 1$, and $3 - f(-0.5)$, following inequalities (9), count the number of times all the three values are positive, and divide this number by 10^3 (the number of trials). This yields an experimental probability (frequency) of the event, $P_{EXP.} = 0.24$ (Figure 4.11, cell *I2*). In column *K* one can see the results of 19 such experiments, each of which included 10^3 trials. Summing all numbers in column *K* and dividing them by 19 yields 0.2505...—even a better approximation to the theoretical result, 0.25, than 0.24 (Figure 4.11, cell *I2*).

9. Loci of Two-Variable Inequalities as Images of Points and Pick's Formula

A remarkable in its simplicity formula for area of a polygon in the lattice plane (a formal analogue of a geoboard—a hands-on environment for learning geometry) with vertices in the points with integer coordinates (lattice points) was found by Pick[3]. The formula says that area *A* of such a polygon can be calculated as follows

$$A = \frac{B}{2} + I - 1$$

where *B* is the number of lattice points on the border and *I* is the number of lattice points in the interior of the polygon .

Pick's formula can be employed to demonstrate yet another way of finding area of the triangle defined by inequalities (10) by using the *GC*. The software can construct a point as a graph of a two-variable inequality. For example, the inequality

$$\sqrt{\varepsilon - |a - a_0|}\sqrt{\varepsilon - |b - b_0|} > 0$$

determines the neighborhood of the point (a_0, b_0) in the coordinate plane (*a*, *b*), where $\varepsilon > 0$ is used to measure the "thickness" of the point (the "size" of the neighborhood). Indeed, as an object in the plane, the ε-neighborhood of the point (a_0, b_0) can be defined through the set

$$\{(a,b) \mid -\varepsilon < a - a_0 < \varepsilon, -\varepsilon < b - b_0 < \varepsilon\}$$

which represents a square formed by the intersection of two mutually perpendicular stripes of width 2ε with the center in (a_0, b_0). Using this graphing technique, one can construct all points with integer coordinates located on the border and in the interior of the triangle shown in Figure 4.8 as the domain in the plane (*a*, *b*) which describes the behavior of the function *f*(*x*) defined in the context of Exploration 4. For example, in order to plot the point (1, 2) that belongs to the border of the triangle, the graph of the inequality

$$\sqrt{0.02 - |x - 1|}\sqrt{\varepsilon - |y - 2|} > 0$$

[3] Georg Alexander Pick (1859-1942)—an Austrian mathematician who worked for more than 40 years in Prague, the capital of what is now Czech Republic.

should be constructed in the plane (x, y) of the *GC* where, as usual, custom variables x and y play the roles of a and b. To plot all points shown in Figure 4.12, one needs to graph 13 inequalities of the above type (nine inequalities corresponding to the border points and four inequalities corresponding the interior points). Counting the points so constructed and using Pick's formula yields

$$A_{TR.} = \frac{9}{2} + 4 - 1 = 7.5 .$$

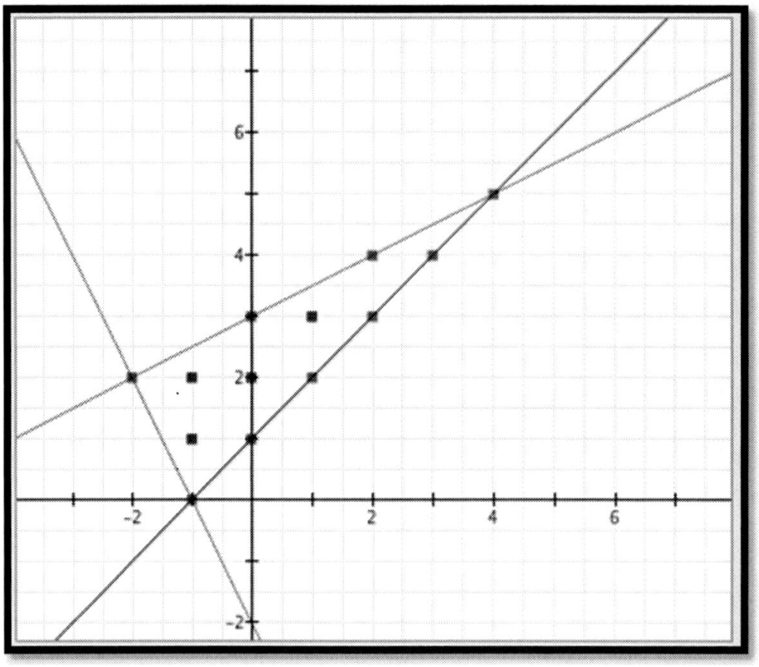

Figure 4.12. Finding area in the plane of parameters using Pick's formula.

Dividing the number 7.5 by the area of rectangle from which parameters a and b may be chosen yields $\frac{1}{4}$, thereby, confirming the results obtained by other methods.

10. Advanced Explorations with Geometric Probabilities

Problem-solving techniques introduced in the previous sections can be applied to finding probabilities of a certain behavior of functions, equations, and systems of equations with random parameters. In some cases, the very regions within which needed behavior occurs may have variable borders, including curvilinear ones. As the first example, consider

Exploration 5. Parameters a and b are chosen at random over the square with side two linear units centered at the origin. Find the probability that the function $f(x) = a^2x + b$ at the points $x = 1$ and $x = -1$ assumes values such that

$$f(1) < 1 , f(-1) > 0 \qquad (12)$$

Reflections. Rewriting inequalities (12) in terms of a and b yields

$$b<1-a^2, \quad b>a^2 \tag{13}$$

In order to construct regions in the plane (a, b) defined by inequalities (13), one has to graph the equations $b=1-a^2$ and $b=a^2$, which define the borders of the region where inequalities (13) hold. This region is the overlap of the two parabolas shown in Figure 4.13 in dark color. Alternatively, the region represents a locus of the two-variable inequality $\sqrt{1-x^2-y}\sqrt{y-x^2}>0$ formulated in custom (for the *GC*) variables x and y.

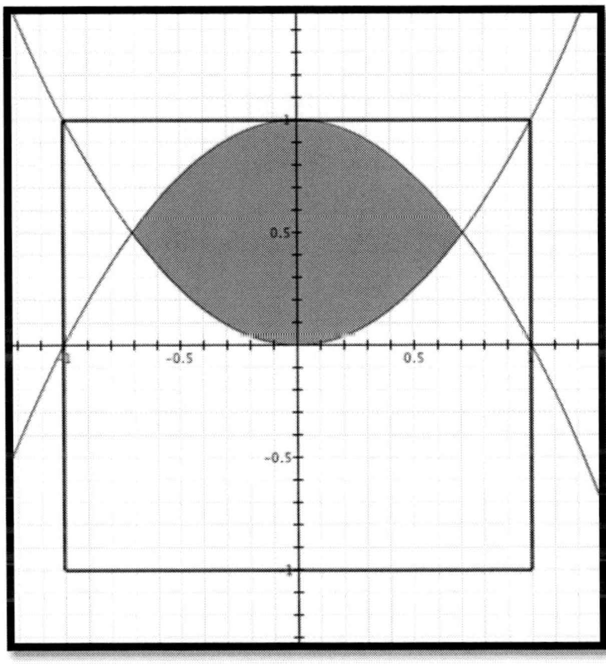

Figure 4.13. Calculating area of the shaded region.

To find the area of the region, one can calculate the integral

$$2\int_0^{a_0}(1-2a^2)da$$

where a_0 is the positive root of the equation $1-a^2=a^2$, i.e., $a_0=1/\sqrt{2}$. Therefore,

$$2\int_0^{1/\sqrt{2}}(1-2a^2)da=2(a-\frac{2}{3}a^3)\Big|_0^{1/\sqrt{2}}=\frac{4}{3\sqrt{2}}.$$

Since the area of a square with side two equals four, the geometric probability p of the event described by inequalities (10) is

$$p = \frac{4}{3\sqrt{2}} \div 4 = \frac{1}{3\sqrt{2}} = 0.2357\ldots .$$

Just like in the case of Exploration 4, this geometric (theoretical) probability can be compared with the corresponding experimental probability using the spreadsheet shown in Figure 4.14. By randomly choosing 10^3 points from the square

$$-1 < a < 1, -1 < b < 1$$

and counting the number of times inequalities (13) hold true, one can arrive at the value of 0.234 as an experimental probability of the event stated in the context of Exploration 5. One can see that the experimental and geometric probabilities found, respectively, through spreadsheet modeling and integration coincide to the accuracy of 0.002.

Exploration 6. Parameter a is chosen at random over the interval (1, 2). What is the probability that both roots of the quadratic equation

$$x^2 - 2(2a - 1)x + 3a - 2 = 0 \qquad (14)$$

are smaller than the number 2?

	A	B	C	D	E	F	G	H	I	J
1	*a*	*b*		*1-f*(1)	*f*(-1)			Exp. Probability		
2	-0.63	0.72		-0.11934807811007	0.329902752241886		234	0.234	1	0.242
3	0.27	-0.59		1.51561994194946	-0.660175201422486				2	0.252
4	-0.09	-0.82		1.81085727565888	-0.828729479855875				3	0.241
5	-0.65	0.84		-0.25971487432192	0.427323322485483				4	0.262
6	0.38	0.90		-0.03841367422954	0.752078043386992				5	0.242
7	0.52	-0.98		1.71383088874493	-1.244915921302230				6	0.247
8	-0.80	-0.88		1.23045104085531	-1.522011590992860				7	0.247
9	0.41	-1.00		1.82879359207418	-1.164480561798160				8	0.243
10	0.87	-0.80		1.04304417535987	-1.560150610220980				9	0.232
11	-0.32	-0.46		1.35640507460954	-0.567195478059547				10	0.228
12	-0.03	0.14		0.85942261930959	0.138835814141749	1			11	0.239
13	-0.24	0.92		0.02288684827670	0.857189220408484	1			12	0.24
14	-0.36	-0.90		1.77605841419870	-1.029668006077490				13	0.208
15	-0.09	-0.54		1.52938618273000	-0.546277450001449				14	0.217
16	0.53	-0.53		1.24159348789499	-0.811276916382206				15	0.241
17	0.61	-0.15		0.77679312968725	-0.527584259049365				16	0.252
18	-0.97	0.62		-0.55396636816253	-0.323873368866352				17	0.24
19	0.06	0.63		0.36464408277087	0.628686244306921	1			18	0.234
20	-0.19	-0.12		1.08482884741496	-0.158974728343687				19	0.246
1001	0.06	0.77		0.23115920838122	0.762010255954597	1				

Figure 4.14. Calculating experimental probability for Exploration 5.

Reflections. First of all, one has to find those values of parameter a that provide equation (14) with real roots residing to the left of the point $x = 2$ on the number line. To this end, one can introduce the function $f(x, a) = x^2 - 2(2a - 1)x + 3a - 2$. As discussed in Chapter 2 in detail (e.g., Exploration 16), in order for both roots of equation (14) to be smaller than the number 2, the following three inequalities

$$f(2,a) > 0,\ 2a - 1 < 2,\ f(2a - 1, a) < 0 \tag{15}$$

are necessary and sufficient.

A graphic representation of inequalities (15) is given in Figure 4.15. Here $2a - 1$ is the x-coordinate of the vertex of parabola $y = f(x, a)$. Formulating inequalities (15) in terms of parameter a yields

$$-5a + 6 > 0,\ a < 1.5,\ 4a^2 - 7a + 3 > 0$$

whence

$$1 < a < \frac{6}{5} \tag{16}$$

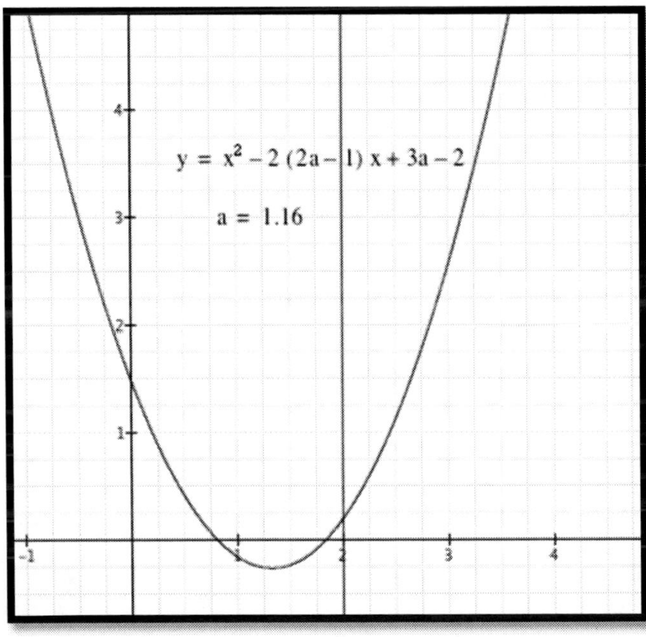

Figure 4.15. Comparing the x-intercepts of the parabola to the point $x = 2$.

According to definition (1), the geometric probability p of the event that equation (14) has both roots smaller than the number 2 is equal to the ratio of the measure of the interval defined by inequalities (16) to the measure of the interval (1, 2); that is,

$$p = \frac{\frac{6}{5} - 1}{2 - 1} = \frac{1}{5}.$$

Exploration 7. Parameters a and b are chosen at random over a square centered at the origin of the plane (a, b). What is the probability that the system of equations

$$x+y=a, \quad x-y=b \tag{17}$$

has a positive solution?

Reflections. Solving the system of equations (17) results in the following solution

$$x=\frac{a+b}{2}, \quad y=\frac{a-b}{2}$$

Therefore, in the plane (a, b) the following inequalities should hold true

$$a+b>0, \quad a-b>0 \tag{18}$$

As the sketch of Figure 4.15 indicates, regardless of the size of square centered at the origin, the probability sought is equal to 0.25 (as the shaded region is bounded by coordinate angle bisectors and, thereby, always covers one-fourth of any square centered at the origin).

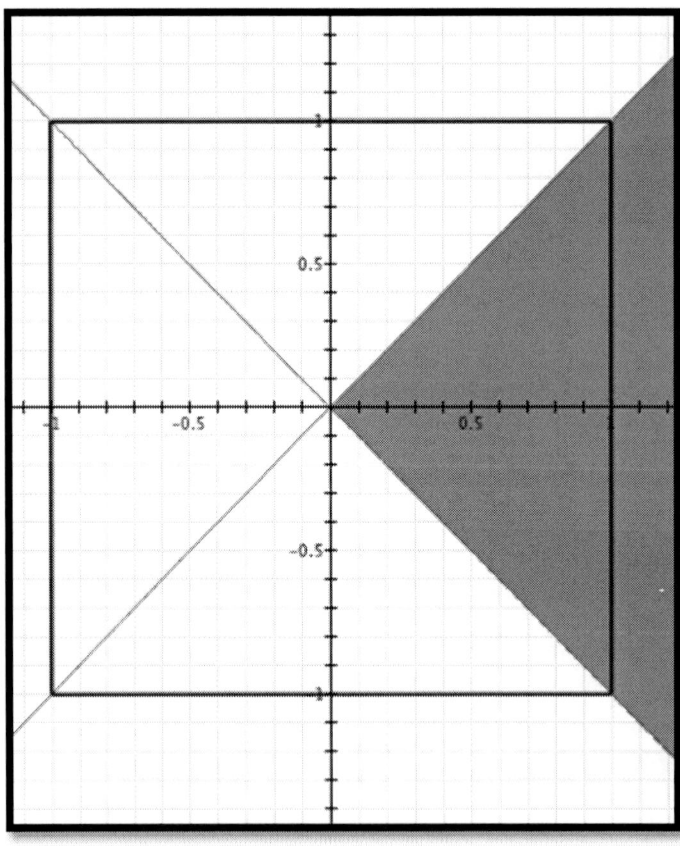

Figure 4.16. The shaded region represents the locus of inequalities (18).

For a more complicated region in the plane of parameters defined by non-linear inequalities its overlapping part with a square centered at the origin depends on the radius of the square, thereby, requiring more involved calculations of geometric probabilities. As an example, consider

Exploration 8. Parameters a and b are chosen at random over a square with side $2R$ centered at the origin. What is the probability that each of the following two systems of simultaneous equations

$$x + y = a, xy = b \tag{19}$$

and

$$x + y = b, xy = a \tag{20}$$

has two real solutions?

Reflections. Let us denote K_{2R} the square from which parameters a and b are chosen. It follows from Viete's theorem (Chapter 2) that any two numbers with the sum a and the product b represent the roots of the quadratic equation $z^2 - az + b = 0$. Therefore, the inequality

$$a^2 - 4b \geq 0 \tag{21}$$

is necessary and sufficient for equations (19) to have real roots. By the same token, in the case of equations (20), such necessary and sufficient condition has the form

$$b^2 - 4a \geq 0 \tag{22}$$

Considering (21) and (22) as simultaneous inequalities, the task is to plot a region in the plane of parameters a and b where both inequalities hold true and then compare this region to the square K_{2R}.

To this end, one can graph thc parabolas $b - \frac{a^2}{4}$ and $a = \frac{b^2}{4}$ in the coordinate plane (a, b). The shaded part of the plane (a, b) shown in Figure 4.17, represents the locus of simultaneous inequalities (21) and (22). Alternatively, one can construct the locus of the inequality $\sqrt{a^2 - 4b}\sqrt{b^2 - 4a} > 0$. One can see that, depending on the value of R, square K_{2R} (an enclosure bounded by the straight lines $a = \pm R$ and $b = \pm R$) may or may not overlap with the upper right part of the locus. As the equation $a = \frac{(\frac{a^2}{4})^2}{4}$ has two real roots, $a = 0$ and $a = 4$, and the corresponding values of b are $b = 0$ and $b = 4$, respectively, the above two parabolas intersect at the point (4, 4) and at the origin. When $R = 4$, the point (4, 4) becomes the top-right corner of the square K_{2R}, the area of which is equal to $4R^2$.

Figures 4.17-4.19 illustrate three mutual arrangements of the square K_{2R} and the locus of inequalities (21) and (22). Thus, three cases need to be considered: $R = 4$, $R < 4$, and $R > 4$. Consequently, geometric probabilities of three events need to be computed: $P(R = 4)$, $P(R < 4)$, and $P(R > 4)$.

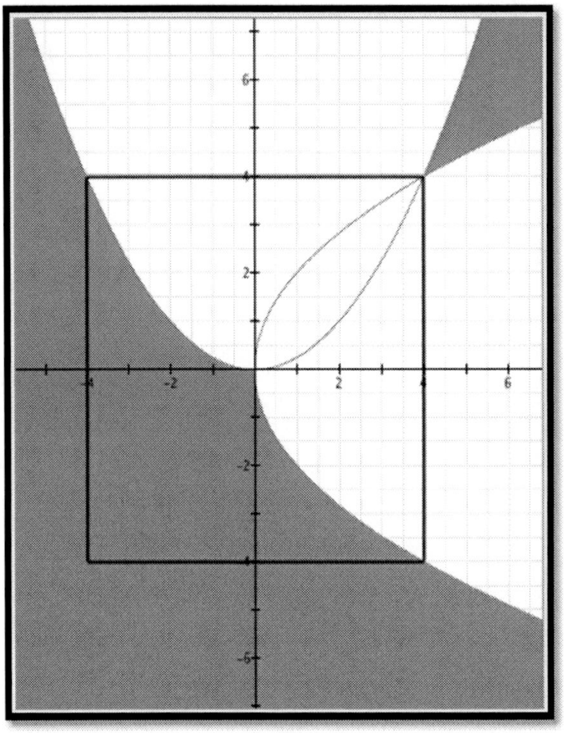

Figure 4.17. The case $R = 4$.

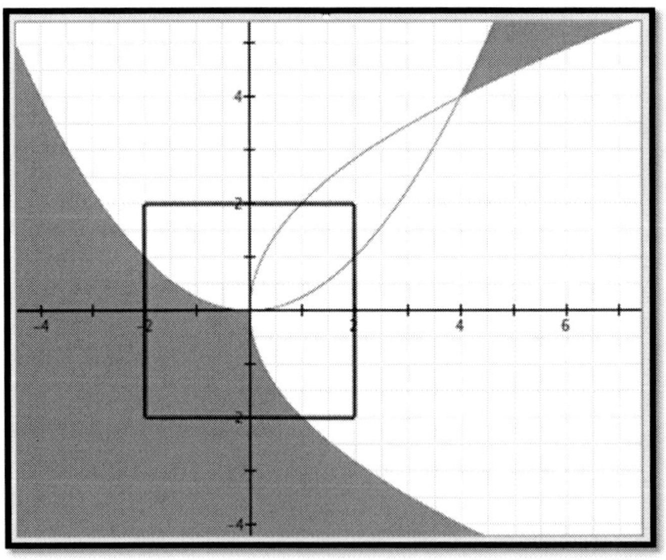

Figure 4.18. The case $R < 4$.

First, consider the case $R = 4$ (Figure 4.17). The area of the shaded region that belongs to the square can be found through integration as twice the area of the curvilinear triangle plus the area of square with side four

$$2\int_{-4}^{0}\frac{a^2}{4}da + 16 = \frac{64}{6} + 16 = \frac{80}{3}.$$

Therefore,

$$P(R = 4) = \frac{80}{3} \div 64 = \frac{5}{12}.$$

Now, consider the case $R < 4$ shown in Figure 4.18. This time, the area of the shaded region that belongs to the square can be found as

$$2\int_{-R}^{0}\frac{a^2}{4}da + R^2 = \frac{R^3}{6} + R^2$$

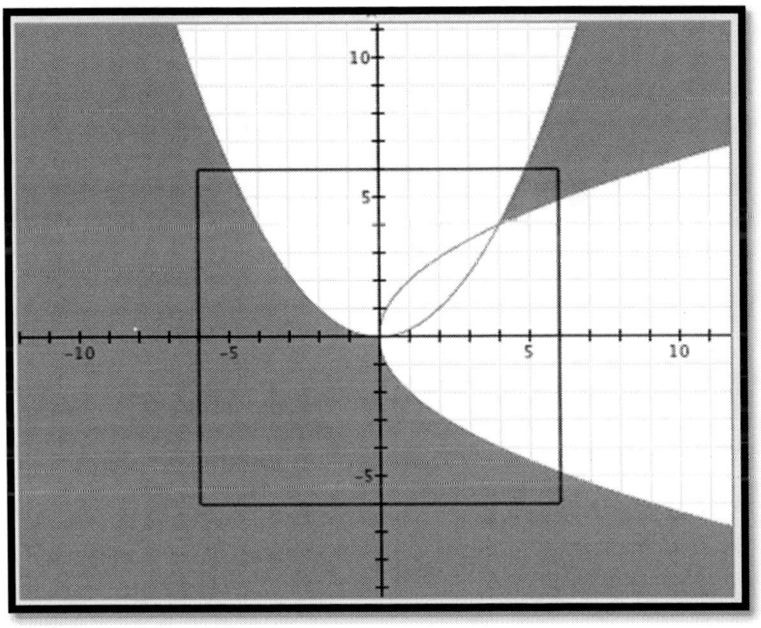

Figure 4.19. The case $R > 4$.

Therefore,

$$P(R < 4) = \frac{\frac{R^3}{6} + R^2}{4R^2} = \frac{R}{24} + \frac{1}{4}$$

One can see that $\lim_{R \to 4} P(R < 4) = \frac{5}{12}$.

Finally, in the case $R > 4$ shown in Figure 4.19 the area of the shaded region that belongs to the square can be calculated by using *Maple* (Figure 4.20):

$$A := 2\cdot\int_{-R}^{0} \frac{a^2}{4}\,da + R^2 + \int_{4}^{R} \left(\frac{a^2}{4} - 2\sqrt{a}\right) da - 3\cdot\int_{-R}^{-2\sqrt{R}} \left(\frac{a^2}{4} - R\right) da$$

$$R^2 + \frac{16}{3} + \frac{2}{3}R^{3/2} + 3R\left(-2\sqrt{R} + R\right)$$

simplify(%)

$$4R^2 + \frac{16}{3} - \frac{16}{3}R^{3/2}$$

Figure 4.20. *Maple* integration in the case $R > 4$.

Therefore,

$$P(R>4) = \frac{4R^2 - \frac{16}{3}R^{\frac{3}{2}} + \frac{16}{3}}{4R^2} = 1 - \frac{4}{3\sqrt{R}} + \frac{4}{3R^2}$$

Note that $\lim\limits_{R\to 4} P(R>4) = 1 - \frac{2}{3} + \frac{1}{12} = \frac{5}{12}$ and $\lim\limits_{R\to\infty} P(R>4) = 1$.

One can also use a spreadsheet to calculate experimental probabilities for different values of R. For example, by randomly selecting 10^3 points (a, b) from the square of side $2R$, the following experimental data can be generated:

$P_{EXP.}(R=2) = 0.329$,
$P_{EXP.}(R=30) = 0.745$, $P_{EXP.}(R=1000) = 0.966$.

At the same time, geometric probabilities found above give the following results:

$P_{GEOM.}(R=2) = 0.333...$,
$P_{GEOM.}(R=30) = 0.758...$, $P_{GEOM.}(R=1000) = 0.958$.

One can recognize quite a good match of the results obtained experimentally (using a spreadsheet) and theoretically (through geometrization).

Exploration 9. Consider the recurrent sequence $f_{n+1} = \sqrt{af_n^2 + b}$, $f_1 = 1$ with random parameters a and b chosen from the square K_{2R} of side $2R$ centered at the origin. Find the probability that the sequence converges to a number smaller than $\frac{1}{\sqrt{2}}$.

Solution. As will be shown in Chapter 6 below, the assumption $|a| < 1$ yields the relationship

$$\lim_{n\to\infty} f_n = \sqrt{\frac{b}{1-a}}$$

Therefore, in order for this limit to be smaller than $\frac{1}{\sqrt{2}}$, that is, in order for the inequality $\sqrt{\frac{b}{1-a}} < \frac{1}{\sqrt{2}}$ hold true, the following three inequalities should be satisfied

$$\frac{b}{1-a} > 0,\ \frac{b}{1-a} < \frac{1}{2},\ |a| < 1 \tag{23}$$

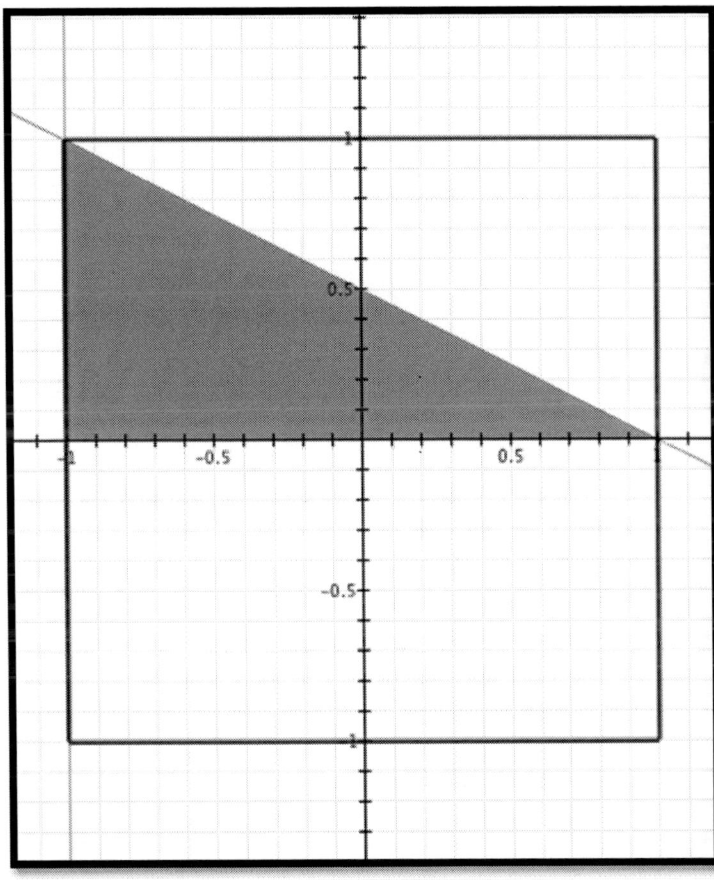

Figure 4.21. Exploration 9: the case $R = 1$.

Inequalities (23) define a region in the plane (a, b) that, as shown in Figures 4.21-4.23, depending on the value of R, the intersection of this region with the square K_{2R} is shaped either as a right triangle ($R = 1$ and $R > 1$) or as a pentagon ($R < 1$). The following three probabilities have to be found: $P(R = 1)$, $P(R > 1)$, and $P(R < 1)$. In all the three cases, the hypotenuse of the right triangle belongs to the straight line

$$b = -\frac{a}{2} + \frac{1}{2} \tag{24}$$

where the inequality $\frac{b}{1-a}<\frac{1}{2}$ turns into the equality $\frac{b}{1-a}=\frac{1}{2}$. Indeed

$$\frac{b}{1-a}=\frac{-\frac{a}{2}+\frac{1}{2}}{1-a}=\frac{1}{2}.$$

Similarly, the horizontal leg of the right triangle belongs to the line $b = 0$. Finally, in all three cases, the area of square K_{2R} is equal to $4R^2$. Equation (24) enables one to calculate the areas of the right triangle and the pentagon in the case $R > 1$ and $R < 1$, respectively.

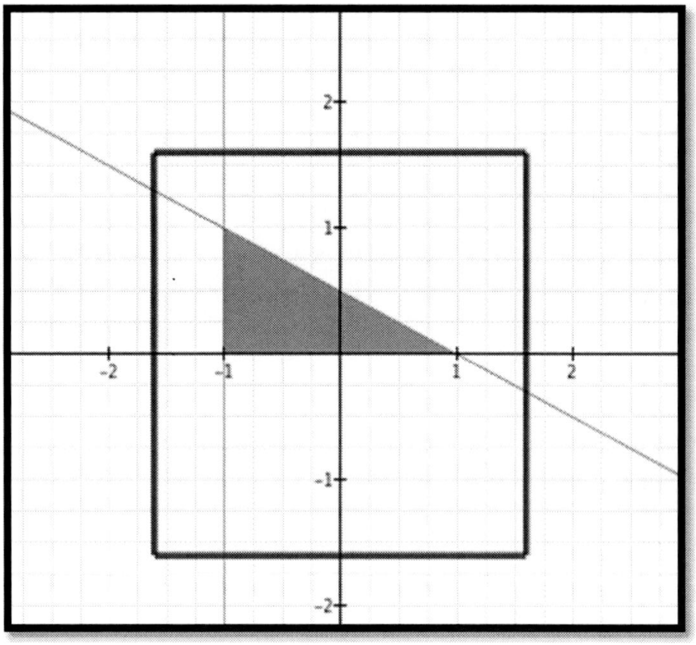

Figure 4.22. Exploration 9: the case $R > 1$.

Consider the case $R = 1$ (Figure 4.21). The shaded triangle has the legs of the lengths one and two linear units. Therefore, as its area is equal to one square unit,

$$P(R=1)=\frac{1}{4}$$

Next, let $R > 1$ (Figure 4.22). The shaded triangle has the legs of the lengths $2R$ and $\frac{R+1}{2}$, the latter being the b-coordinate of the point on the straight line $b=-\frac{a}{2}+\frac{1}{2}$ through which the vertical line $a = -R$ passes. Therefore, the area of this triangle is equal to $\frac{R(R+1)}{2}$ whence

$$P(R > 1) = \frac{R(R+1)}{2} \div (4R^2) = \frac{R+1}{4R}.$$

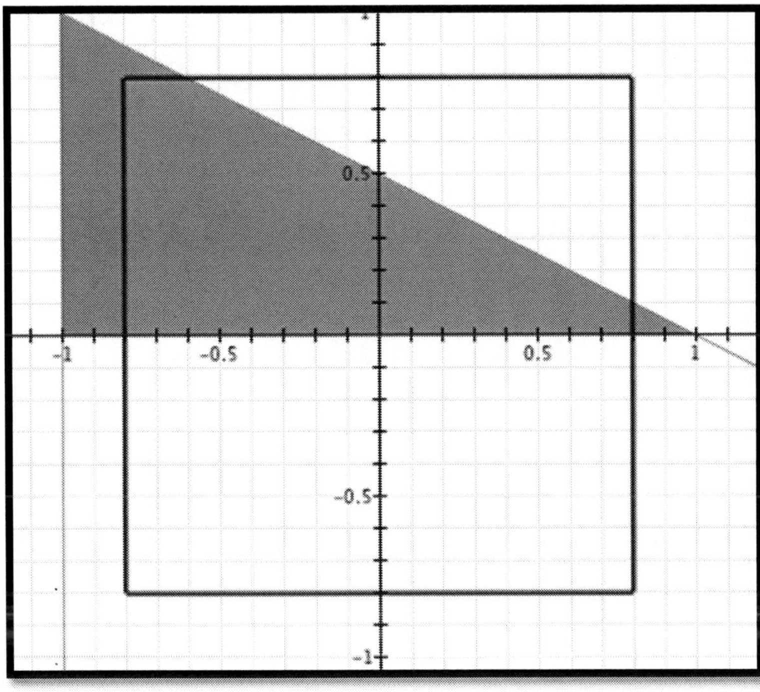

Figure 4.23. Exploration 9: the case $R < 1$.

Finally, consider the case $R < 1$ (Figure 4.23). The shaded pentagon can be represented as the union of the rectangle with side lengths $2R$ and $\frac{1-R}{2}$, the latter being the b-coordinate of the point on the straight line $b = -\frac{a}{2} + \frac{1}{2}$ through which the vertical line $a = R$ passes, and the trapezoid with the bases of the lengths $2R$ and $1 - R$. In order to find the latter length one has to find the a-coordinate of the point on the straight line $b = -\frac{a}{2} + \frac{1}{2}$ through which the horizontal line $b = R$ passes. To this end, one has to find a from the equation $R = -\frac{a}{2} + \frac{1}{2}$ yielding $a = 1 - 2R$.

	A	B	C	D	E	F	G	H	I	J	K	L
1	a	b		b/(1-a)	1/2-b/(1-a)	1-a	a+1			Exp. Probability		
2	-0.47	0.18		0.12374584	0.38	1.47	0.53	1	451	0.45	1	0.448
3	-0.06	-0.32		-0.3055544	0.81	1.06	0.94				2	0.462
4	0.37	-0.07		-0.1103718	0.61	0.63	1.37				3	0.416
5	0.47	-0.28		-0.5340348	1.03	0.53	1.47				4	0.431
6	-0.04	0.15		0.14051676	0.36	1.04	0.96	1			5	0.421
7	0.21	-0.12		-0.1456639	0.65	0.79	1.21				6	0.443
8	0.21	-0.13		-0.1694602	0.67	0.79	1.21				7	0.436
9	0.20	0.13		0.1573404	0.34	0.80	1.20	1			8	0.417
10	0.31	-0.34		-0.4893467	0.99	0.69	1.31				9	0.474
11	-0.28	-0.38		-0.2957281	0.80	1.28	0.72				10	0.42
12	-0.26	-0.36		-0.2878229	0.79	1.26	0.74				11	0.434
13	0.32	-0.24		-0.355505	0.86	0.68	1.32				12	0.419
14	0.35	0.08		0.11938153	0.38	0.65	1.35	1			13	0.443
15	0.30	-0.02		-0.0226061	0.52	0.70	1.30				14	0.448
16	0.24	0.04		0.05660352	0.44	0.76	1.24	1			15	0.447
17	0.15	0.44		0.51631101	-0.02	0.85	1.15				16	0.434
18	0.03	0.23		0.23301738	0.27	0.97	1.03	1			17	0.432
19	-0.06	0.11		0.1012765	0.40	1.06	0.94	1			18	0.431
20	0.16	-0.17		-0.1966582	0.70	0.84	1.16				19	0.451
1001	-0.27	0.10		0.07691147	0.42	1.27	0.73	1				

Figure 4.24. Calculating experimental probability for Exploration 9.

Then, one has to find the distance between the points on the a-axis with the coordinates $a = 1-2R$ and a = -R. Next, one has to find the height $\frac{3R-1}{2}$ of the pentagon as the difference

$$R - \frac{1-R}{2} = \frac{2R-1+R}{2} = \frac{3R-1}{2}.$$

Therefore, the area of the rectangle is equal to $R(1 - R)$ and the area of the trapezoid is equal to the average of its bases multiplied by the height

$$\frac{2R+1-R}{2} \cdot \frac{3R-1}{2} = \frac{(R+1)(3R-1)}{4}.$$

Finally, the area of the pentagon is equal to

$$R(1-R) + \frac{(R+1)(3R-1)}{4} = \frac{4R(1-R)+(R+1)(3R-1)}{4} = \frac{-R^2+6R-1}{4}$$

whence

$$P(R<1) = \frac{-R^2+6R-1}{4} \div (4R^2) = -\frac{1}{16} - \frac{1}{16R^2} + \frac{3}{8R}.$$

Note that

$$\lim_{R\to 1} P(R<1) = \lim_{R\to 1}\left(-\frac{1}{16} - \frac{1}{16R^2} + \frac{3}{8R}\right) = -\frac{1}{16} - \frac{1}{16} + \frac{3}{8} = \frac{1}{4}.$$

To conclude note that one can use a spreadsheet to find an experimental probability of the event stated in the context of Exploration 9. Let $R = 0.5$. Choosing 10^3 random points from a unit square centered at the origin that belong to the shaded region (Figure 4.23). The spreadsheet of Figure 4.24 demonstrates that how such approach was implemented in the case $R = 0.5$ yielding the value of the experimental probability $P_{EXP}(R = 0.5) = 0.45$. By the same token, the probability $P(R < 1)$ in the case $R = 0.5$ is equal to

$$P(R = 0.5) = -\frac{1}{16} - \frac{1}{16 \cdot 0.25} + \frac{3}{8 \cdot 0.5} = \frac{7}{16} = 0.44 .$$

In much the same way, other cases of R can be explored through a spreadsheet-based computational experiment.

11. ACTIVITY SET

1. A machine changed a half dollar coin into quarters, dimes, and nickels. Assuming that there are equally likely to get any combination of the coins, find the probability that there were no nickels in the change.
2. In the context of Exploration 8, what is the largest value of $R < 4$ for which the inequality $P(R < 4) \leq 1/3$ holds?
3. Parameters a and b are chosen at random over the rectangle which vertices are (-1, 0), (-1, 8), (7, 0), and (7, 8). What is the probability that the function $f(x) = ax + b$ at the points $x = 1$, $x = -2$, and $x = -0.5$ assumes values such that $f(1) < 11$, $f(-2) < 2$ and $f(-0.5) > 0.5$?
4. Parameters a and b are chosen at random over the triangle which vertices are (-4, 0), (0, 8) and (8, 0). What is the probability that the linear function $f(x) = ax + b$ at the points $x = 1$, $x = -2$ and $x = -0.5$ assumes values such that $2 < f(1) < 8$, $f(-2) > -4$ and $f(-0.5) < 5$?
5. Parameters a and b are chosen at random over the circle of radius 2 centered at the point (2, 5). What is the probability that the linear function $f(x) = ax + b$ at the points $x = 1$, $x = -1$, $x = 3$ and $x = -3$ assumes values such that $f(1) < 9$, $f(-1) > 5$, $f(3) > 9$ and $f(-3) < -3$?
6. Parameters a and b are chosen at random over the circle $a^2 + b^2 = R^2$. What is the probability that the system of simultaneous equations $ax + by = 1, x^2 + y^2 = 1$ has solutions such that $x > y$?
7. Parameters p and q are chosen at random over the circle $p^2 + q^2 = R^2$. What is the probability that the equation $x^3 + px + q = 0$ has three real roots?
8. Parameters b and c are chosen at random over the square of the side length four centered at the origin. What is the probability that the roots x_1 and x_2 of the equation $x^2 + bx + c = 0$ satisfy the inequalities $x_1 < c < x_2$?
9. Parameters b and c are chosen at random over the square of the side length R centered at the origin. What is the probability that the equation $x^2 + bx + c = 0$ has three real roots?

Chapter 5

COMBINATORIAL EXPLORATIONS

One of the first and foremost duties of the teacher is not to give his students the impression that mathematical problems have little connection with each other.
— Pólya (1973, p. 15)

1. INTRODUCTION

Combinatorics is one of the oldest branches of discrete mathematics whose content can be traced back to the 16th century when games of chance played an important role in the life of a society. The need for a theory of such games stimulated the development of new mathematical concepts and counting techniques. At the basic level, being relevant to students' experiences, combinatorial problems include counting ways to check out five books out of ten books put on display, or buy five donuts out of ten types offered by the grocery store. Stemming from real-life context, combinatorial counting provides strong motivation for the learning of mathematics. In contemporary classroom, such counting can be enhanced by the use of different technology tools.

Historically, discrete mathematics was not considered a separate curriculum strand until the last decade of the 20th century, and its many concepts and ideas were incorporated into the classic areas of mathematics such as geometry, number theory, and algebra. However, due to the integration of computers into pre-college mathematics curricula discrete mathematics was elevated to the status of a separate standard, though only for the upper grades (National Council of Teachers of Mathematics, 1989). One may note that unlike the 1989 document, in its more recent version—Principles and Standards for School Mathematics (National Council of Teachers of Mathematics, 2000)—discrete mathematics does not appear as a separate standard. Nevertheless, the Technology Principle asserts, "many discrete mathematics topics take on new importance in the contemporary mathematics classroom"(ibid, p.27).

One of the recommendations by the Conference Board of the Mathematical Sciences (2001) for the preparation of teachers concerns the need to become familiar with ideas, methods, and applications associated with difference equations—recursive formulation of discrete concepts. Difference equations are used as mathematical models of discrete dynamical systems found in applications to radio engineering, communications, and computer architecture. This chapter will show how a spreadsheet can generate integers with combinatorial meaning by numerically modeling difference equations that describe the concepts of combinatorics. The simplest combinatorial concept is that of permutations. It is described by a non-linear difference equation depending on one integer variable. Other combinatorial concepts considered in this chapter include combinations (without and with repetitions), sums of perfect powers, partitions into distinct parts (known as Stirling numbers of the second kind), and permutations with rises (known as Eulerian numbers). These concepts depend on two integer variables and can be expressed through partial difference equations (linear and otherwise) subject to boundary conditions. In general, activities described in the chapter are designed to help teachers experience the contiguity of different concepts of discrete mathematics and appreciate their integrity—a significant source in the development of mathematical knowledge.

2. Two Ways of Defining Permutations

The concept of *permutations* describes all possible arrangements of a set of objects, each of which contains every object once with two such arrangements differing only in the order of their elements. A typical situation for combinatorial analysis is to focus on the *number* of permutations rather than on creating the permutations. However, for a small number of objects, the process of creating permutations is instructive. As an example, consider the set of three letters, {A, B, C}. We can say that the "word" ABC is a permutation of the three letters. Other permutations include ACB, BAC, BCA, CAB, and CBA. One can recognize a system used in permuting the letters. Indeed, all permutations were put into three groups depending on the choice of the first letter: choose A as the first letter and permute the other two letters, B and C; choose B as the first letter and permute A and C; and, finally, choose C as the first letter and permute A and B.

Denoting $P(n)$ the number of permutations of n objects, the above example of counting permutations by actually creating them can be described as follows

$$P(3) = P(2) + P(2) + P(2) = 3 \cdot P(2).$$

In order to generalize from the case $n = 3$, consider

Library Display Problem. How many ways can n books—B_1, B_2, ..., B_n—be put on display in different orders?

To answer this question, let the books form the set $\{B_1, B_2, ..., B_n\}$ of n objects. There exist $P(n - 1)$ permutations with B_1 as the first object, $P(n - 1)$ permutations with B_2 as the first object, ..., $P(n - 1)$ permutations with B_n as the first object. Therefore, one can count the number of permutations $P(n)$ of n objects (or, in our case, the number of orders in which the books can be put on display) through what may be called the rule of repeated sum:

$$P(n) = \underbrace{P(n-1) + P(n-1) + ... + P(n-1)}_{n \text{ addends}}$$

or

$$P(n) = n \cdot P(n-1) \tag{1}$$

In order for equation (1) to be complete, one has to define $P(0)$—the number of permutations of zero objects. To this end note that $P(1) = 1$—one object can be permuted in one way only—and, therefore, equation (1) implies the equality

$$P(0) = 1 \tag{2}$$

Considered jointly, relations (1) and (2) represent a recursive definition of the permutations of n objects, $P(n)$, and may be considered as the simplest example of a difference equation with a variable coefficient subject to an initial condition.

One can note that recursive definition (1)-(2) does not allow for an immediate answer to the library display problem. A closed formula for $P(n)$ is needed. Such a formula follows from the recursive definition. Indeed,

$$P(n) = n \cdot P(n-1) = n \cdot (n-1) \cdot P(n-2) = n \cdot (n-1) \cdot (n-2) \cdot P(n-3)$$
$$= \ldots = n \cdot (n-1) \cdot (n-2) \cdot \ldots \cdot 2 \cdot 1 = n!$$

Therefore,

$$P(n) = n! \tag{3}$$

Note that relations (2) and (3) are in compliance with the definition of the permutation of zero objects, $0! = 1$.

	A	B	C
1	n	$P(n)$	$n!$
2	0	1	1
3	1	1	1
4	2	2	2
5	3	6	6
6	4	24	24
7	5	120	120
8	6	720	720
9	7	5040	5040
10	8	40320	40320
11	9	362880	362880
12	10	3628800	3628800
13	11	39916800	39916800
14	12	479001600	479001600

Figure 5.1. Permutations generated through recursive and closed formulas.

Another way to derive formula (3) is to use

The Rule of Product. If the object B_1 can be selected in m_1 ways and if, following the selection of B_1, the object B_2 can be selected in m_2 ways, …, and, finally, after the selection of $n - 1$ objects, the object B_n can be selected in m_n ways, then the permutation $[B_1, B_2, \dots, B_n]$ can be created in $m_1 m_2 \dots m_n$ ways.

Therefore, a book to be displayed first can be selected in n ways, a book to be displayed second can be selected in $n - 1$ ways, . . . , and, finally, for the last position we then have one choice only to select a book. Once again, $P(n) = n \cdot (n-1) \cdot \ldots \cdot 2 \cdot 1 = n!$.

The spreadsheet pictured in Figure 5.1 generates numbers $P(n)$ through both recursive (column B) and closed (column C) definitions. One may note that the first three values of $P(n)$ coincide with the first three Fibonacci[1] numbers defined by the linear difference equation $F(n+1) = F(n) + F(n-1)$, where $F(0) = F(1) = 1$, but then increase much faster due to their non-linear growth. Another difference is that, unlike Fibonacci numbers, permutations do not exhibit any interesting patterns to be mentioned. The reason to mention Fibonacci numbers in the context of combinatorics is due to their interesting connection to combinations that will be discussed below.

3. Counting Permutations of Repeated Objects

Consider now the case of counting permutations of repeated objects. How many ways can one permute $n + r$ elements in the set $\Omega = \left\{ \underbrace{A, A, \dots, A}_{rA's}, B_1, B_1, \dots, B_n \right\}$, where r elements are identical? To answer this question let P^r_{n+r} denote the number of permutations sought. For each of the P^r_{n+r} permutations, r letters A can be permuted in $r!$ ways. By the rule of product, the number of permutations on the set Ω is equal to the product $P^r_{n+r} \cdot r!$. On the other hand, the number of permutations of $r + n$ objects is equal to $(r + n)!$. Therefore, $P^r_{n+r} \cdot r! = (n+r)!$ whence

$$P^r_{n+r} = \frac{(n+r)!}{r!} \qquad (4)$$

In particular, the number of permutations of letters in the word ALABAMA can be found as $P^3_{3+4} = \frac{7!}{4!} = 210$. Formula (4) can be extended to the case of several repeated objects in a set to be permuted. For example, the number of permutations of letters in the word MISSISSIPPI is equal to $\dfrac{11!}{\underbrace{(4!)}_{four\ I's}\ \underbrace{(4!)}_{four\ S's}\ \underbrace{(2!)}_{two\ P's}} = 34{,}650.$

---

[1] Leonardo Fibonacci (c. 1170-1250), also known as Leonardo of Pisa—an Italian mathematician who introduced the Hindu-Arabic number system to Europe.

4. Defining Combinations through a Partial Difference Equation

Modeling of many real-life situations deals with the concept of combination—the arrangement of objects in a set when their order is immaterial. To clarify, consider

Award Selection Problem. There is a pool of n people nominated for r different awards, $r \leq n$. Assuming that a person may not be given more than one award, determine how many ways selections for the awards can be made.

To put this inquiry in a more abstract context, consider a set of n objects. Any choice of r objects from among n is considered as an r-combination of n objects, with two such combinations being different only if they differ in composition. The problem, thereby, is to find the total number of r-combinations of n objects. Hereafter, this number, as a function of two integer variables n and r, will be denoted $C(n,r)$. There is a closed formula

$$C(n,r) = \frac{n!}{r!(n-r)!} \tag{5}$$

which defines $C(n, r)$ and, thereby, gives a solution to the award selection problem[2]. However, the Conference Board of the Mathematical Sciences (2000) emphasized the importance of developing teachers' ability to solve problems of that type through recursive reasoning, rather than through direct application of formula (5). Earlier, the National Council of Teachers of Mathematics (1989) came up with a similar recommendation: "instruction should emphasize combinatorial reasoning as opposed to the application of analytic formulas for permutations and combinations" (p. 179). Combinatorial reasoning enables one to introduce $C(n, r)$ through a recursive definition (already used to introduce permutations)—the method of reducing a problem to an analogous problem involving a smaller number of objects. This, in turn, requires the use of the general rule of combinatorics, known as

The Rule of Sum. If an object A can be selected in m ways and an object B can be selected in n ways, then either A or B can be selected in $m + n$ ways.

For example, if publisher A offers five mathematics textbooks priced below \$100, and if publisher B offers three mathematics textbooks priced below \$100, then such a mathematics textbook from either publisher can be selected in eight ways.

In order to apply the rule of sum to the award selection problem, one can choose a person, named, say, Amy, and then divide all combinations of people selected for awards into two groups. The first group includes Amy; the second group does not include Amy. The number of combinations in the first group is equal to $C(n-1,r-1)$; indeed, as Amy is always included in a combination, one needs to create $(r - 1)$-combinations out of $n - 1$ people. The number of combinations in the second group is equal to $C(n-1,r)$; indeed, as Amy is not included in a combination, this time one needs to create r-combinations out of $n - 1$ people. The rule of sum yields the partial difference equation

[2] The definition $0! = 1$ mentioned earlier allows to apply formula (5) when $n = r$.

$$C(n,r) = C(n-1,r-1) + C(n-1,r) \quad (6)$$

Because $C(n,0)$—the number of combinations of n people zero at a time—has no combinatorial meaning, one can define $C(n,0) = 1$ for all $n \geq 0$. To demonstrate that this definition makes sense, note that there exist n ways of choosing one person from the pool of n people, that is, $C(n,1) = n$. Thus, it follows from equation (6) that

$$C(n,0) = C(n+1,1) - C(n,1) = n+1-n = 1.$$

Defining $C(0,r) = 0$ for $r > 0$completes boundary conditions for equation (6) as follows

$$C(0,r) = 0, r \geq 1; C(n,0) = 1, n \geq 0; \quad (7)$$

Finally, in order to derive formula (5), one can solve the award selection problem through counting permutations of repeated objects. To this end, any selection of r nominees out of n people can be recorded through an n-letter word consisting of r letters Y and $n - r$ letters N, where Y and N mean nomination and non-nomination, respectively. For example, the diagram

{Amy, Bob, Carol, Don, Eileen, Fred, George}➔{YNYYNNY}

may be used to indicate that Amy, Carol, Don, and George were nominated for four awards, while Bob, Eileen, and Fred were not nominated. Therefore, the total number of ways to select four nominees out of seven people can be found as the number of permutations of letters in a word with four Y's and three N's. Extending formula (4) to the case of two repeated letters, this number can be found as $\dfrac{7!}{\underbrace{(4!)}_{\text{four Y's}}\ \underbrace{(3!)}_{\text{three N's}}} = 35$. In the general case, n people can be nominated (selected) for r different awards in $\dfrac{n!}{r!(n-r)!}$ ways, thus formula (5). In the case (mentioned in the introduction) of checking out five books out of ten books, the total number of combinations is equal to $\dfrac{10!}{5!\cdot 5!} = 252$.

5. NUMERICAL MODELING AS A WAY OF MAKING MATHEMATICAL CONNECTIONS

Figure 5.2 shows the spreadsheet, which numerically models equation (6) subject to boundary conditions (7). Observing numbers generated by the spreadsheet, one can associate equation (6) with its "graph" shown in the diagram of Figure 5.3. The diagram represents the following simple rule: the combined content of the non-shaded cells is equal to the content of the shaded cell. This rule will also be used to develop and prove several combinatorial identities involving combinations *C*(*n*, *r*).

	A	B	C	D	E	F	G	H	I	J	K	L
1	*r**n*	0	1	2	3	4	5	6	7	8	9	10
2	0	1	1	1	1	1	1	1	1	1	1	1
3	1	0	1	2	3	4	5	6	7	8	9	10
4	2	0	0	1	3	6	10	15	21	28	36	45
5	3	0	0	0	1	4	10	20	35	56	84	120
6	4	0	0	0	0	1	5	15	35	70	126	210
7	5	0	0	0	0	0	1	6	21	56	126	252
8	6	0	0	0	0	0	0	1	7	28	84	210
9	7	0	0	0	0	0	0	0	1	8	36	120
10	8	0	0	0	0	0	0	0	0	1	9	45
11	9	0	0	0	0	0	0	0	0	0	1	10
12	10	0	0	0	0	0	0	0	0	0	0	1

Figure 5.2. Combinations without repetitions.

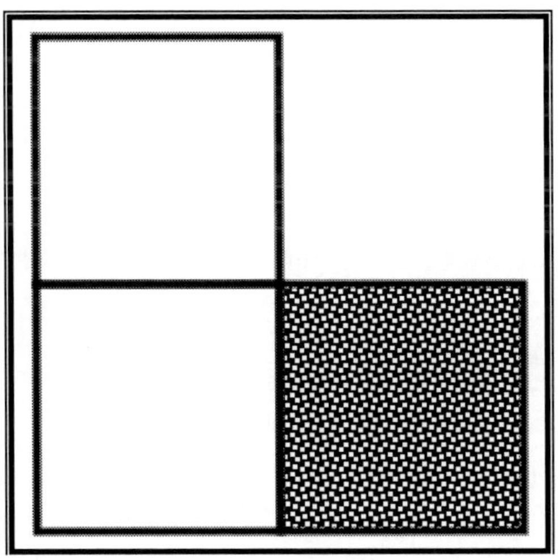

Figure 5.3. The "graph" of combinations.

One can note that the array of numbers shown in Figure 5.2 represents a rectangular form of Pascal's triangle[3] the nth column of which (located above zeros) is the sequence of the

[3] Blaise Pascal—a French mathematician, physicist, and philosopher of the 17th century, one of the founders of probability theory.

coefficients of $x^{n-r}y^r$ in the expansion of the binomial $(x + y)^n$, i.e., the sequence $\{C(n,r)\}_{r=0}^{n}$. For example, $(x+y)^3 = x^3 + 3x^2y + 3xy^2 + y^3$ in the case n = 3 (cell *E1*) and the coefficients in this expansion are the numbers located in the range *E2:E5*. In general, the following binomial expansion

$$(x+y)^n = C(n,0)x^n + C(n,1)x^{n-1}y + C(n,2)x^{n-2}y^2$$
$$+ ... + C(n,n-1)xy^{n-1} + C(n,n)y^n$$

holds true.

Interpreting the array of combinations $C(n, r)$ as a re-oriented Pascal's triangle, one can conjecture the following property of combinations

$$C(n, r)=C(n, n-r), \quad 0 \le r \le n \tag{8}$$

Contextually, relation (8) demonstrates that the number of ways of choosing r people from the pool of n people is equal to that of choosing $n - r$ people from n people. Indeed, one can pair off each r-combination of n people with the $(n - r)$-combination of $n - r$ remaining people. In this pairing, different r-combinations determine different $(n - r)$-combinations, and vice versa. In other words, the choice of r people from n people is equivalent to the choice of $n - r$ remaining people. Put another way, when a coin is tossed n times, the number of outcomes with exactly r heads is equal to the number of outcomes with exactly $n - r$ heads. For example, when $n = 3$ the corresponding sample space is the set {HHH, TTT, HHT, HTH, THH, TTH, THT, HTT} with the subsets {HHT, HTH, THH} and {TTH, THT, HTT} of equal cardinality. Therefore, this sample space can be represented through the quadruple (1, 3, 3, 1)—the binomial coefficients in the expansion of $(x + y)^3$. In fact, Pascal came across his famous triangle as a way of recording sample spaces of experiments with n coins for different values of n (Kline, 1985).

Identity (8) can also be proved by the direct application of formula (5). Indeed, according to formula (5)

$$C(n,n-r) = \frac{n!}{(n-(n-r))!(n-r)!} = \frac{n!}{r!(n-r)!} = C(n,r).$$

Adding numbers in the corresponding columns beginning from column *B* and row *2* (Figure 5.2) yields the following sequence of integers (the last integer is the sum of the numbers in the range *L2:L12*)

1, 2, 4, 8, 16, 32, 64, 128, 256, 512, 1024

which represents the sequence $\{2^n\}_{n=0}^{10}$. In other words,

$$C(n,0) + C(n,1) + C(n,2) + ... + C(n,n) = 2^n.$$

Also, within the array of numbers shown in Figure 5.2, one can discover Fibonacci numbers as the sums of its elements located along the so-called shallow diagonals with the end points $C(n, 0)$ and $C(n - r, r)$ where, given n, either $r = \frac{n}{2}$ or $r = \frac{n+1}{2}$. For example, when $n = 6$ the sum of the numbers in the corresponding shallow diagonal (with the end points in cells *H2* and *E5*) is equal to 13—the 7th Fibonacci number. Indeed,

$$C(6, 0) + C(5, 1) + C(4, 2) + C(3, 3) = 1 + 5 + 6 + 1 = 13.$$

In general, the following remarkable identity connecting combinations to Fibonacci numbers holds true:

$$C(n,0) + C(n-1,1) + C(n-2,2) + ... + C(n-r,r) = F_{n+1} \quad (9)$$

where $r = \frac{n}{2}$ when n is an even number and $r = \frac{n+1}{2}$ otherwise.

Identity (9) can be provided with an interesting combinatorial interpretation. To this end, consider the following problematic situation that, in particular, can be used for the introduction of Fibonacci numbers.

Painting Problem. Buildings of different number of stories are given and one has to paint them with a fixed color in such a way that no two consecutive stories are painted with it. How many ways of such painting of one, two, three, four, etc.–storied buildings are possible? Note: not painting a building at all is considered a special case of painting as in that case the main condition of not having consecutive stories painted is satisfied.

Solution. Consider a five-storey building. One way of painting is not to paint it at all. This can be done in one way only and can be described as $C(6, 0)$. Another possibility is to paint just one storey. This can be done in $C(5, 1)$ ways as, by definition, the number of ways to select one storey out of five stories is $C(5, 1)$. The next possibility is to paint two stories. This can be done in $C(4, 2)$ ways. Indeed, when two adjacent stories in a four-storey building are painted, one can add an additional storey to separate them; otherwise, one can add a storey somewhere in between two painted ones. In both cases, one returns back to a five-storey building painted according to the rules. Painting three stories can be done in $C(3, 3)$ ways. As $C(3, 3) = 1$, we have a completely painted three-storey building. Adding two additional stories to separate the painted ones yields a five-storey building painted according to the rules. Therefore, the painting of a five-storey building can be done in $C(6, 0) + C(5, 1) + C(4, 2) + C(3, 3) = 13$ ways and $13 = F_{6+1}$.

On the other hand, all five-storey buildings painted according to the rule can be put in two groups, depending on whether the fifth storey is painted or not (Figure 5.4). When the fifth storey is not painted, then the fourth storey may be painted and, therefore, the number of buildings with the non-painted fifth storey is equal to the number of ways a four-storey building can be painted. When the fifth storey is painted, the fourth storey may not be painted and, therefore, the number of buildings with the painted top storey is equal to the number of ways a three-storey building can be painted. Using the notation PAINT(n) to represent the number of ways an n-storey building can be painted, one has PAINT(5) = PAINT(4) +

PAINT(3)—the form of the relation through which Fibonacci numbers develop. Figure 5.4 shows that PAINT(5) = 8 + 5 = 13—the 7th Fibonacci number.

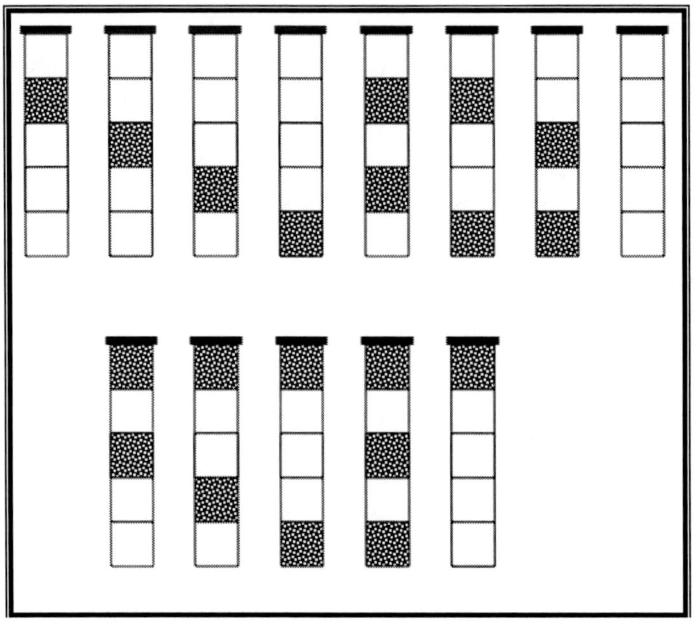

Figure 5.4. The painting problem in the case $n = 5$.

In general, there exist $C(n - r, r)$ ways to paint an $(n - 1)$-storey building so that exactly r non-adjacent stories are painted with one color. Indeed, $C(n - r, r)$ represents the number of ways to choose r objects (e.g., stories) out of $n - r$ objects (stories), where $0 \le r \le n$. In order to separate r painted stories, one needs to insert $(r - 1)$ additional stories. In that way, a $(n - r)$-storied building turns into the building with $(n - r) + (r - 1) = n - 1$ stories in which r non-adjacent stories are painted. Therefore, a $(n - 1)$-storey building can be painted in $\sum_{r=0}^{INT(\frac{n+1}{2})} C(n-r,r)$ ways, where $INT(x)$ is the largest integer smaller than or equal to x.

In particular, a one-storey building can be painted in

$$\sum_{r=0}^{INT(\frac{2+1}{2})} C(2-r,r) = C(2,0) + C(1,1) = 1 + 1 = 2$$

ways, a two-storey building can be painted in

$$\sum_{r=0}^{INT(\frac{3+1}{2})} C(3-r,r) = C(3,0) + C(2,1) + C(1,2) = 1 + 2 + 0 = 3$$

ways, a three-storey building can be painted in

$$\sum_{r=0}^{INT(\frac{4+1}{2})} C(4-r,r) = C(4,0) + C(3,1) + C(2,2) = 1+3+1 = 5$$

ways, and a four-storey building can be painted in

$$\sum_{r=0}^{INT(\frac{5+1}{2})} C(5-r,r) = C(5,0) + C(4,1) + C(3,2) + C(2,3) = 1+4+3+0 = 8$$

ways. Note that 2, 3, 5, and 8 are consecutive Fibonacci numbers.

6. Combinations with Repetitions

So far, the notion of a combination included different elements. This notion can be extended to include elements not all of which are different from each other; that is, one can talk about combinations with *repeated* elements. With this in mind, consider

Billiard Balls Problem. Suppose that each of the r identical billiard balls has to be colored using one of the n given colors. How many different colorings are possible?

This situation is close to the award selection problem since the order in which the balls have to be colored is immaterial. On the other hand, in the latter problem combinations with repetitions are allowed (e.g., all the balls may be colored with one color). Problems of this type lead to the so-called combinations with repetitions. Just like in the case of combinations, the billiard ball problem can be formulated in a more abstract form as follows: *There are* n *different categories (e.g., colors) that can be assigned to* r *identical objects (e.g., billiard balls). How many ways can this be done?*

Let x_i represent the number of objects assigned the i-th category, $x_i \geq 0$, $i = 1, 2, \ldots, n$. Then the sum $x_1 + x_2 + \ldots + x_n$ represents the total number of objects assigned either category and, thereby, it is equal to r. In other words, the problem is to find all possible partitions of r into n non-negative integers by solving the Diophantine equation

$$x_1 + x_2 + \ldots + x_n = r.$$

For example, $\{x_1 = r, x_2 = \ldots = x_n = 0\}$is one such solution to this equation. In terms of the billiard balls problem, this solution means that all the balls are colored the same. In the case (mentioned in the introduction) of buying five donuts out of ten types we have the equation $x_1 + x_2 + \ldots + x_{10} = 5$ and the choice $\{x_1 = 5, x_2 = \ldots = x_{10} = 0\}$ means that all five donuts chosen are of the same kind. As will be shown below, this choice (solution) is one out of 2002 total; i.e., one can assign ten categories to five objects in 2002 ways.

Hereafter, the number of ways to assign n categories to r identical objects (different combinations or r identical billiard balls colored with n colors) will be denoted $\overline{C}(n,r)$. Just like in the case of $C(n, r)$, a difference equation for $\overline{C}(n,r)$ can be constructed by using recursive reasoning. To this end note that when one category out of n is fixed (let it be the n-th category), it is possible to either use it in a combination or not. So, all combinations can be divided into two groups depending on whether the n-th category is used or not. If it is not used in combinations, there exist $\overline{C}(n-1,r)$ ways to assign n – 1 categories to r objects. If the n-th category is used in combinations, it is assigned to at least one object and, thereby, there exist $\overline{C}(n,r-1)$ ways to assign n categories to the remaining r – 1 objects. Applying the rule of sum yields the partial difference equation

$$\overline{C}(n,r)=\overline{C}(n-1,r)+\overline{C}(n,r-1) \tag{10}$$

Now, one has to define the boundary conditions $\overline{C}(n,0)$ for all $n \geq 1$ and $\overline{C}(1,r)$ for all $r \geq 1$. Knowing the values of $\overline{C}(n,r)$ on the boundaries $r = 0$ and $n = 1$ would allow one to fill up the first row and the first column of the corresponding spreadsheet when modeling equation (10) numerically. To this end, using the billiard balls problem as a frame of reference, let x_i denote the number of the balls colored with the i-th color, $x_i \geq 0$, $i = 1, 2, \ldots, n$. Then $x_1 + x_2 + \ldots + x_n = r$ and $\overline{C}(n,r)$ is the number of solutions to this equation. When $r = 0$ the only solution to the equation $x_1 + x_2 + \ldots + x_n = 0$ regardless of n is $\{x_1 = x_2 = \ldots = x_n = 0\}$. The case $n = 1$ yields $x_1 = r$ and, once again, there exists one solution only. This leads to the following boundary conditions

$$\overline{C}(n,0)=1,\ n \geq 1 \text{ and } \overline{C}(1,r)=1,\ r \geq 1 \tag{11}$$

for equation (10). The spreadsheet shown in Figure 5.5 generates combinations with repetitions defined jointly by relations (10) and (11).

	A	B	C	D	E	F	G	H	I	J	K
1	r\n	1	2	3	4	5	6	7	8	9	10
2	0	1	1	1	1	1	1	1	1	1	1
3	1	1	2	3	4	5	6	7	8	9	10
4	2	1	3	6	10	15	21	28	36	45	55
5	3	1	4	10	20	35	56	84	120	165	220
6	4	1	5	15	35	70	126	210	330	495	715
7	5	1	6	21	56	126	252	462	792	1287	2002
8	6	1	7	28	84	210	462	924	1716	3003	5005
9	7	1	8	36	120	330	792	1716	3432	6435	11440
10	8	1	9	45	165	495	1287	3003	6435	12870	24310
11	9	1	10	55	220	715	2002	5005	11440	24310	48620
12	10	1	11	66	286	1001	3003	8008	19448	43758	92378

Figure 5.5. Combinations with repetitions.

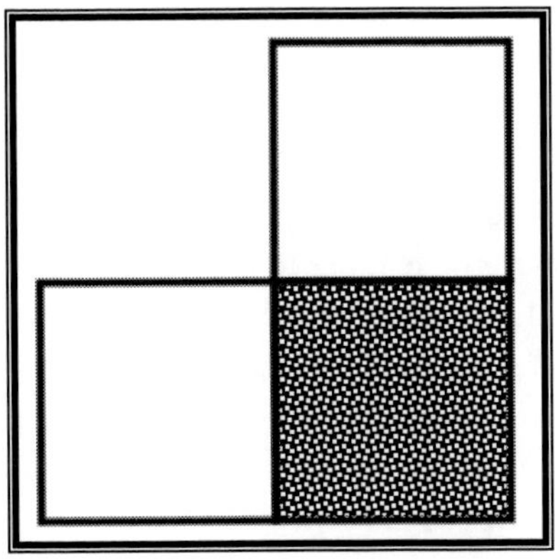

Figure 5.6. The "graph" of combinations with repetitions.

Numerical data generated by the spreadsheet, just like in the case of combinations without repetitions, makes it possible to introduce the "graph" of equation (10)—a visual image of combinations with repetitions. Considering cells *F6, F5*, and *E6* (Figure 5.5), one can see that 70 = 35 + 35. Such a relation among three neighboring cells (forming a flipped L shape) remains true for the whole template and it is expressed through the diagram of Figure 5.6. The diagram reads as follows: the combined content of the two non-shaded cells is equal to the content of the shaded cell.

Finally, like in the case of formula (5), a closed formula for $\overline{C}(n,r)$ can be derived. To this end, note that any choice of coloring r billiard balls using n colors can be recorded through a ($r + n - 1$)-letter word consisting of r letters Y and $n - 1$ letters S, where letter Y indicates the choice of a particular color and letter S is used to separate colors. For example, the diagram

{Six billiard balls; Colors: Blue, Green, Red, Yellow} →
$\{\underbrace{YY}_{BLUE}\ SS\ \underbrace{Y}_{RED}\ S\ \underbrace{YYY}_{YELLOW}\}$

may be used to indicate that two balls are colored Blue, green color is not used (there is no letter Y between the first and the second letters S), one ball is colored red, and three balls are colored yellow. Therefore, the total number of ways to color six billiard balls using four colors can be found as the number of permutations of letters in a word with six Y's and three S's (one needs three separators for four colors). Once again, extending formula (4) to the case of two repeated letters, the number $\overline{C}(4, 6)$ can be found as follows

$$\overline{C}(4,6) = \frac{(6+4-1)!}{\underbrace{(6!)}_{six\ Y's}\ \underbrace{(3!)}_{three\ S's}} = 84\,.$$

In the general case, the number of ways n different categories can be assigned to r identical objects can be found through the closed formula

$$\bar{C}(n,r) = \frac{(r+n-1)!}{(r!)\cdot(n-1)!} \tag{12}$$

Indeed, one needs $n-1$ letters S to separate n categories and r letters Y to represent r, perhaps repeated, objects, thereby, leading to a word with $n-1$ letters S and r letters Y. The right-hand side of formula (12) represents the number of permutations of the letters in such a word, where $\underbrace{YY...Y}_{r\ letters}\underbrace{SS...S}_{n-1\ letters}$ is one such permutation.

7. Combinatorial Identities and Mathematical Induction Proof

One of the recommendations of the Conference Board of the Mathematical Sciences (2001) for the preparation of teachers includes the need for "courses that develop deep understanding of mathematics they will teach" (p. 7) and demonstrate "what it means to write a formal proof" (p. 14). It has been observed through the teaching of a capstone course that within spreadsheet-supported learning environments the teachers enjoy dealing with mathematical formalism and develop proficiency in writing proofs. Indeed, "designing a useful spreadsheet requires flexible ability to express numerical relationships in algebraic notation" (ibid, pp. 127-128) and encourages the emergence of "deep questions [by the teachers] about the appropriate role of skill in algebraic manipulation" (p. 124). Therefore, it is useful to combine an informal spreadsheet exploration with a formal mathematical demonstration. Below, a spreadsheet will be used as support system in developing mathematical induction proof of relations connecting integers with combinatorial meanings. Such relations are often referred to as combinatorial identities. In addition, *Maple* will be used to support symbolic computations in the context of proving combinatorial identities.

7.1. Identities Involving *C(n, r)*

Discovering, explaining, and generalizing computer-generated numerical patterns constitute an important aspect of using technology in mathematics education. A number of such patterns can be discovered within the spreadsheet of Figure 5.2. For example, consider combinations of four cells arranged in the shapes shown in the diagrams of Figure 5.7 and Figure 5.8. In a symbolic form, these diagrams can be described, respectively, as follows

$$C(n,r) = C(n-1,r-1) + C(n-2,r-1) + C(n-2,r) \tag{13}$$

and

$$C(n,r) = C(n-1,r) + C(n-2,r-1) + C(n-2,r-2) \tag{14}$$

Let $(n, r) = (10, 4)$ in relation (13). Then, $210 = 84 + 56 + 70$ (as can be seen in Figure 5.2). How does one know that this is a correct equality? The correctness of this particular equality is easy to verify by adding the three numbers, either mentally or using a calculator. However, in order to demonstrate that relation (13) is true for all n and r and, thereby, represents an identity, one has to work with symbols by using the rules of algebra. Nowadays, this demonstration can be done either by paper-and-pencil or using software capable of symbolic computations. Just like in the case of verifying with a calculator that the sum 84 + 56 + 70 is equal to 210, one can use *Maple* to verify that the right-hand side of (13) is identical to $C(n, r)$ as defined through formula (5). Of course, there should be a right balance of when to use technology, both in numeric and symbolic contexts. We argue that in the case of relation (13) the use of *Maple* is didactically warranted, as one needs to carry out quite cumbersome algebraic manipulations associated with factorials[4].

Figure 5.9 shows how the *Maple*-based simplification of the right-hand side of relation (13) yields $C(n, r)$ expressed through formula (5). Similarly, relation (14) can be verified by using *Maple* (Figure 5.10).

Identity (13) described by Figure 5.7 can be evolved by the repeated application of formula (5) or of the "graph" shown in Figure 5.3 to the following form

$$\begin{aligned} C(n,r) &= C(n-1,r-1)+C(n-2,r-1)+\ldots \\ &+C(n-(n-r+1),r-1) \end{aligned} \tag{15}$$

Note that the last term in the right-hand side of identity (15) can be written in the form $C(r-1, r-1)$ and therefore, a seemingly missing term $C(n-(n-r+1), r) = C(r-1, r)$ is equal to 0. A geometric representation of identity (15) is shown in Figure 5.11.

To clarify, consider a numeric example. When $(n, r) = (8, 4)$ we have $C(8, 4) = 70$ (Figure 5.2, cell *J6*). Connecting cell *J6* to the cells in row *5* yields the equality

$$70 = 35 + 20 + 10 + 4 + 1$$

which is a special case of identity (15) for $n = 8$ and $r = 4$.

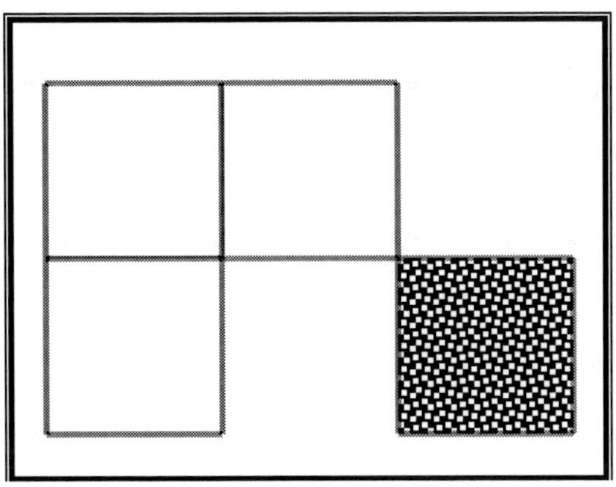

Figure 5.7. Geometric representation of identity (13).

[4] See also a footnote made in Chapter 1 in the context of proving Proposition 20.

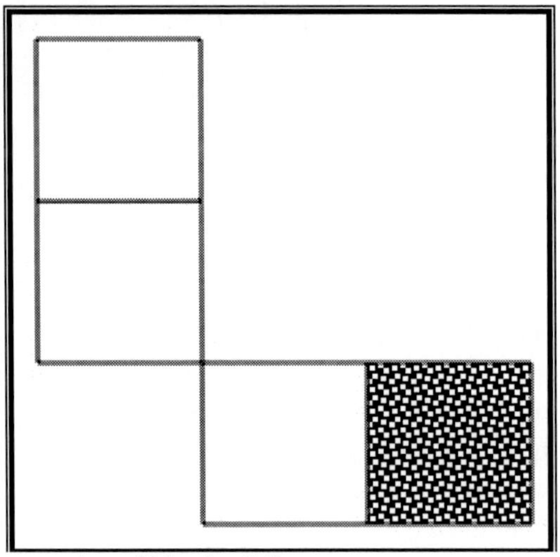

Figure 5.8. Geometric representation of identity (14).

$$C(n,r) := \frac{n!}{r!\cdot(n-r)!}$$

$$(n,r) \rightarrow \frac{n!}{r!\,(n-r)!}$$

$$A1 := C(n-1, r-1)$$

$$\frac{(n-1)!}{(r-1)!\,(n-r)!}$$

$$A2 := C(n-2, r-1)$$

$$\frac{(n-2)!}{(r-1)!\,(n-1-r)!}$$

$$A3 := C(n-2, r)$$

$$\frac{(n-2)!}{r!\,(n-2-r)!}$$

$$A4 := A1 + A2 + A3$$

$$\frac{(n-1)!}{(r-1)!\,(n-r)!} + \frac{(n-2)!}{(r-1)!\,(n-1-r)!} + \frac{(n-2)!}{r!\,(n-2-r)!}$$

$$simplify(A4)$$

$$\frac{n!}{r!\,(n-r)!}$$

Figure 5.9. *Maple*-based proof of identity (13).

In much the same way, identity (14) described by Figure 5.8 can be evolved to the following form

$$\begin{aligned} C(n,r) &= C(n-1,r) + C(n-2,r-1) + C(n-3,r-2) \\ &\quad + \ldots + C(n-(r+1), r-r) \end{aligned},$$

that is,

$$\begin{aligned} &C(n,r) = C(n-1,r) + C(n-2,r-1) + C(n-3,r-2) + \ldots \\ &+C(n-r-1,0) \end{aligned} \qquad (16)$$

$$C(n, r) := \frac{n!}{r! \cdot (n-r)!}$$

$$(n, r) \rightarrow \frac{n!}{r!\,(n-r)!}$$

$$A1 := C(n-1, r)$$

$$\frac{(n-1)!}{r!\,(n-1-r)!}$$

$$A2 := C(n-2, r-1)$$

$$\frac{(n-2)!}{(r-1)!\,(n-1-r)!}$$

$$A3 := C(n-2, r-2)$$

$$\frac{(n-2)!}{(r-2)!\,(n-r)!}$$

$$A4 := A1 + A2 + A3$$

$$\frac{(n-1)!}{r!\,(n-1-r)!} + \frac{(n-2)!}{(r-1)!\,(n-1-r)!} + \frac{(n-2)!}{(r-2)!\,(n-r)!}$$

$$simplify(A4)$$

$$\frac{n!}{(n-r)!\,r!}$$

Figure 5.10. *Maple*-based proof of identity (14).

To clarify, once again, consider a numeric example using the spreadsheet of Figure 5.2. When $(n, r) = (9, 3)$ we have $C(9, 3) = 84$ (cell *K5* in Figure 5.2). Connecting cell *K5* to the sequence of diagonally evolving cells (starting from cell *J5*) yields $84 = 56 + 21 + 6 + 1$—a special case of identity (16) for $n = 9$ and $r = 3$.

The integration of a spreadsheet into the context of proving combinatorial identities provides an excellent medium for teachers not only to practice but also appreciate the essence of mathematical induction proof. As was already mentioned in Chapter 1, a proof by mathematical induction significantly differs from reasoning by induction in the sense that the former is a rigorous argument and the latter is not. Typically, mathematical induction is applied to proving a statement depending on one variable, like identity (15), in which only the first coordinate of the function $C(n, r)$ changes as the right-hand side evolves. As mentioned in section 2 of Chapter 1, the proof begins with establishing a base clause for induction and then continues with the demonstration of the transition from n to $n+1$. This transition can be supported by its diagrammatic representation in a spreadsheet environment (Figure 5.11).

To prove identity (15) using the method of mathematical induction by n note that when $n = 1$ identity (15) is true as it turns into the equality $C(1, r) = C(0, r - 1)$, that is, into the obvious statement $0 = 0$. This constitutes the base of mathematical induction. Assuming that identity (15) is true (this assumption is expressed geometrically through the diagram at the top of Figure 5.11), one has to demonstrate (either algebraically or geometrically) that it remains true when n is replaced by $n + 1$ (see the bottom part of Figure 5.11), that is,

$$C(n+1,r) = C(n,r) + C(n-1,r-1) + C(n-2,r-1) + ... + C(r-1,r-1)$$

Indeed, using equation (6) and (assumed to be true) identity (15), one can write

$$C(n+1,r) = C(n,r) + C(n,r-1)$$
$$= C(n,r-1) + C(n-1,r-1) + C(n-2,r-1) + ... + C(r-1,r-1).$$

Geometrically, when, using the "graph" of combinations (Figure 5.3), the shaded cell in the bottom part of the diagram of Figure 5.11 is replaced by two cells, its neighbor to the left can be replaced by the chain of non-shaded cells from the top part of the diagram of Figure 5.11, thereby yielding the diagram shown at the bottom of Figure 5.11. Identity (15) has been proved both algebraically and geometrically.

The method of mathematical induction can also be used to prove identity (16). However, unlike identity (15), as the right-hand side of (16) evolves, both coordinates of $C(n, r)$ change. In this case, one has to carry out induction by both n and r. As shown in Figure 5.12 (the top part of which represents the "graph" of identity (15)), the diagram evolves diagonally, that is, in both directions (up and to the left).

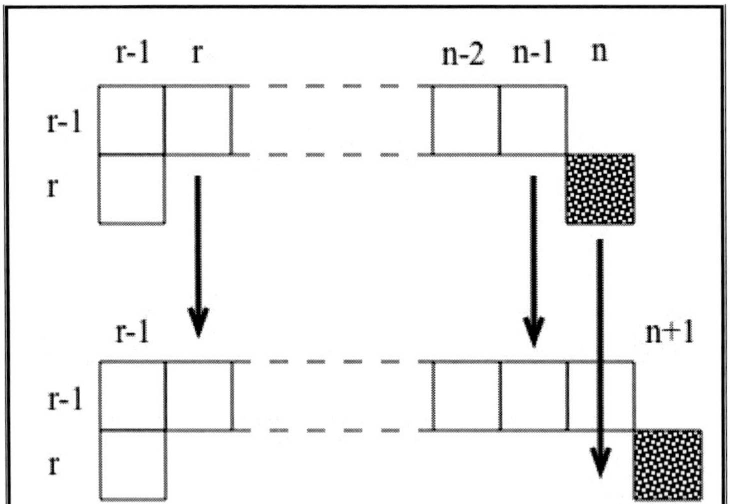

Figure 5.11. The "graph" of inductive transfer from n to $n + 1$ in the case of identity (15).

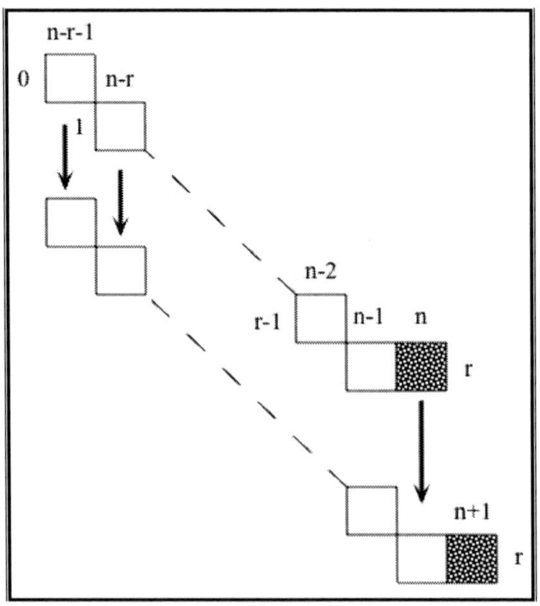

Figure 5.12. Inductive transfer in the case of identity (16).

When $r = 1$ identity (16) is true as the relation $C(n, 1) = C(n-1, 1) + C(n-2, 0)$ is equivalent to the equality $n = (n-1) + 1$ and the latter is true for all n. This constitutes the base of mathematical induction. Assuming that (16) is true (this assumption is expressed geometrically through the diagram at the top of Figure 5.12), one has to show that it remains true when the pair (n, r) is replaced by the pair $(n + 1, r + 1)$, that is,

$$C(n+1,r+1) = C(n,r+1) + C(n-1,r) + C(n-2,r-1)$$
$$+ \ldots + C(n-r-1,0)$$

Indeed, using equation (6) and (assumed to be true) identity (16) one can write

$$C(n+1,r+1) = C(n,r+1) + C(n,r)$$
$$= C(n,r+1) + C(n-1,r) + C(n-2,r-1) + \ldots + C(n-r-1,0).$$

Geometrical representation of the proof, i.e., transition from (n, r) to $(n + 1, r + 1)$, is shown in Figure 5.12. Indeed, when, using the "graph" of combinations (Figure 5.3), the shaded cell in the bottom part of the diagram of Figure 5.12 is replaced by two cells, its neighbor to the left can be replaced by the diagonal-like chain of non-shaded cells from the top part of the diagram of Figure 5.12, thereby yielding the bottom part of Figure 5.12. Identity (16) has been proved both algebraically and geometrically.

7. 2. Identities Involving $\bar{C}(n,r)$

Consider the diagram of Figure 5.13 that results from the application of the "graph" of $\bar{C}(n,r)$ (Figure 5.6) to itself. The four-cell shape means that adding the contents of three non-shaded cells yields the content of the shaded cell. Indeed, by substituting the top cell (Figure 5.6) with two (diagonally connected) cells, results in the shape presented in Figure 5.13.

This shape can evolve further up to assume the form shown in the left part of the diagram of Figure 5.14. In order to translate the geometric image into a symbolic form, note that in the spreadsheet environment of Figure 5.5, any step to the left (right) decreases (increases) the first coordinate by one, and any step down (up) increases (decreases) the second coordinate by one also. This observation results in the following symbolic representation of the left part of the diagram of Figure 5.14 in which the dark cell represents $\bar{C}(n,r+k)$:

$$\bar{C}(n,r) + \bar{C}(n-1,r+1) + \bar{C}(n-1,r+2) + \ldots + \bar{C}(n-1,r+k)$$
$$= \bar{C}(n,r+k) \quad (17)$$

Figure 5.14 can also support the development of proof of identity (17) using the method of mathematical induction by r. Indeed, the base clause for identity (17) is a trivial one, because the chain of the non-shaded cells on the left of the diagram reduces to nothing when $k = 0$ and, therefore, the two cells on the right merge into one cell. The inductive transfer can be interpreted as follows: the assumption that the sum of the non-shaded cells is equal to the shaded cell (the left part of the diagram), remains true when k is replaced by $k+1$.

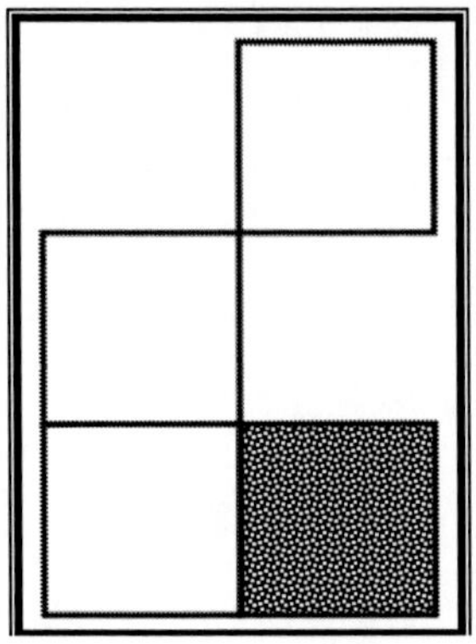

Figure 5.13. Applying the "graph" of $\overline{C}(n,r)$ to itself.

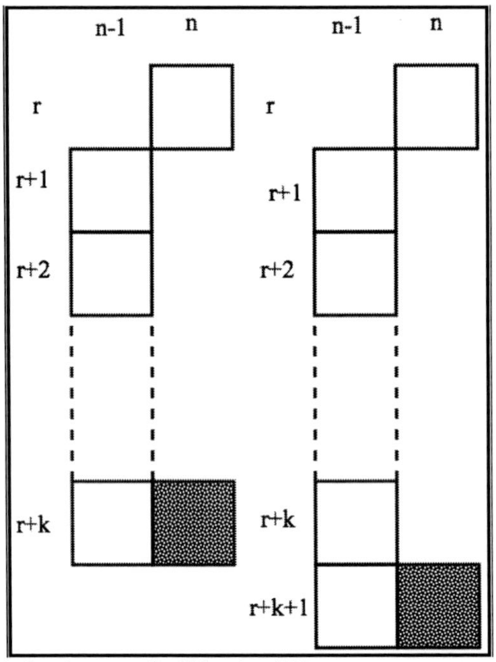

Figure 5.14. Mathematical induction proof of identity (17).

Now, one can carry out the mathematical induction proof of identity (17) by variable r in a symbolic form. When $k = 0$, a true equality $\overline{C}(n,r)=\overline{C}(n,r)$ constitutes the base of mathematical induction. Assuming identity (17) to be true for a certain value of k, results in the following transformation of the sum of $k + 1$ terms

$$\overline{C}(n,r)+\overline{C}(n-1,r-1)+...+\overline{C}(n-1,r+k)+\overline{C}(n-1,r+k+1)$$
$$=\overline{C}(n,r+k)+\overline{C}(n-1,r+k+1)=\overline{C}(n,r+k+1)$$

This demonstrates the transition from $r + k$ to $r + k + 1$ and, thereby, completes the mathematical induction proof of identity (17).

In much the same way, one can develop another evolving shape by successive application of the "graph" of $C(n, r)$ to the lower-left cell (Figure 5.6). In doing so, one can construct a diagram shown in the upper part of Figure 5.15, and then symbolically describe the diagram in the following form

$$\begin{aligned}&\overline{C}(n,r)+\overline{C}(n+1,r-1)+\overline{C}(n+2,r-1)+...+\overline{C}(n+k,r-1)\\&=\overline{C}(n+k,r)\end{aligned} \tag{18}$$

Once again, the diagram of Figure 5.15 can support the development of mathematical induction proof of identity (18) by *n*. As above, with $k = 0$ the chain of the non-shaded cells reduces to nothing which results in the shaded cell turning into the far-left cell. This observation constitutes the base of mathematical induction. The inductive transfer can be interpreted as follows: in virtue of the diagram of Figure 5.6, the assumption that the sum of the left cell and the upper chain of *k* cells is equal to the lower right cell, remains true when *k* is replaced by $k + 1$.

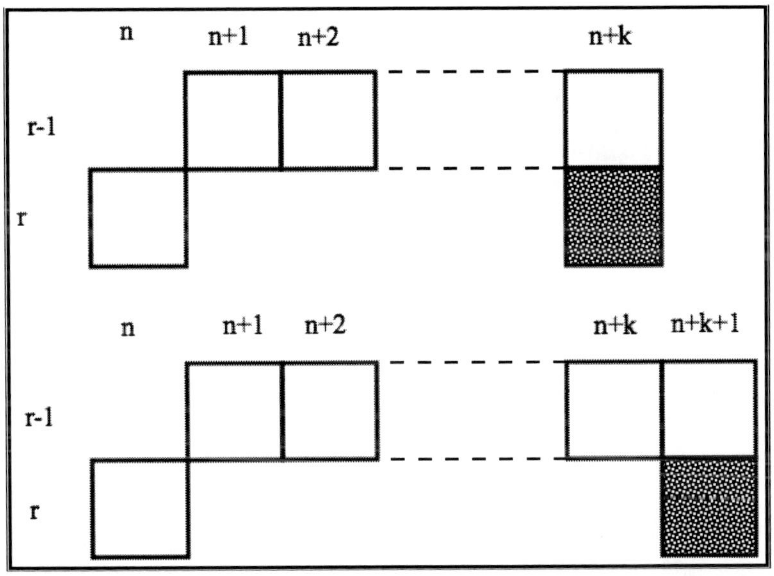

Figure 5.15. Mathematical induction proof of identity (18).

To carry out the proof symbolically with respect to *n*, one can note that when $k=0$ (the base clause), identity (18) is obviously true. Assuming identity (18) to be true for a certain value of *k* (numbers *n* and *r* are fixed) results in the following symbolic representation of the inductive transfer (transition from $n + k$ to $n + k + 1$):

$$\begin{aligned}&\overline{C}(n,r)+\overline{C}(n+1,r-1)+\overline{C}(n+2,r-1)\\&+...+\overline{C}(n+k,r-1)+\overline{C}(n+k+1,r-1)\\&=\overline{C}(n+k,r)+\overline{C}(n+k+1,r-1)=\overline{C}(n+k+1,r).\end{aligned}$$

Identity (18) has been proved both algebraically and geometrically.

8. CONNECTING NUMBERS WITH DIFFERENT COMBINATORIAL MEANING

Another interesting computer-enhanced activity is to establish connections between integers with combinatorial meaning generated by two different spreadsheets. To this end, consider the case of conjecturing a link between $C(n,r)$ and $\overline{C}(n,r)$ by comparing data generated by the spreadsheets of Figure 5.2 and Figure 5.5. Both numerical arrays look rather similar and the task is to establish connections among their elements. To this end, comparing, respectively, the numbers in cells *F7*, *F5*, *D8* of the spreadsheet of Figure 5.5 to those in *K7*, *I5*, *J8* of the spreadsheet of Figure 5.2, one can recognize the following equalities

$$\overline{C}(5,5) = C(9,5), \overline{C}(5,3) = C(7,3), \overline{C}(3,6) = C(8,6).$$

A simple conjecture stemming from the limited evidence could be to notice the sameness of the second coordinates of the functions $\overline{C}(n,r)$ and $C(n,r)$. But is this always true? For example, one can discover that $\overline{C}(8,5) \neq C(9,5)$—a counterexample that defeats the above conjecture. Thus, in general, $\overline{C}(n_1,r)$ and $C(n_2,r)$ do not coincide. What else can be said about the last three equalities? How are the first coordinates of the functions $\overline{C}(n,r)$ and $C(n,r)$ connected within these equalities? One can check to see that the second coordinates can be linked to the first coordinates through the following relations

$$5 = 9 - 5 + 1,\ 3 = 7 - 5 + 1,\ 6 = 8 - 3 + 1,$$

which can be rewritten as

$$9 = 5 + 5 - 1,\ 7 = 5+ 3 - 1,\ 8 = 3 + 6 - 1.$$

Generalizing from the three special cases yields the identity

$$\overline{C}(n,r) = C(n+r-1,r) \tag{19}$$

In order to give combinatorial proof of identity (19), one can group the elements in each r-combination by category so that the elements of the first category go first, then go the elements of the second category, then those of the third category, and so on. Next, we enumerate the elements according to their positions in the combination except that we increase these numbers by one for the elements of the second category, by two for elements of the third category, and by $n - 1$ for elements of the nth category. In this way, each r-combination with repetition, $\overline{C}(i,r)$, $i = 1, 2, \ldots, n$, determines r-combination without repetition C $(j, r), j = 1, 2,..., n + r - 1$. This implies identity (19).

In order to give an analytical proof of identity (19), one can use formulas (5) and (12). According to formula (5)

$$C(n+r-1,r)=\frac{(n+r-1)!}{r!(n-1)!}.$$

On the other hand, according to formula (12)

$$\overline{C}(n,r)=\frac{(r+n-1)!}{r!(n-1)!}.$$

This completes the proof of identity (19).

Formulas (19) and (8) make it possible to establish the property of a diagonal-like symmetry for combinations with repetitions. Indeed,

$$\overline{C}(n,r)=C(n+r-1,r)=C(n+r-1,n-1)$$
$$=C(r+1+n-1-1,n-1)=\overline{C}(r+1,n-1)$$

That is,

$$\overline{C}(n,r)=\overline{C}(r+1,n-1) \qquad (20)$$

Note that identity (20) can be expressed as follows: the equations

$x_1 + x_2 + \ldots + x_n = r$ and $x_1 + x_2 + \ldots + x_{r+1} = n - 1$

have the same number of non-negative integer solutions. For example, when $n = r = 2$ the equations $x_1 + x_2 = 2$, and $x_1 + x_2 + x_3 = 1$ both have, respectively, the following three solutions: (1, 1), (2, 0), (0, 2), and (1, 0, 0), (0, 1, 0), (0, 0, 1). One can be encouraged to explore a few more special cases of n and r as a way of testing identity (20).

9. Partitioning Problems and Recursive Reasoning

Consider the following modification of difference equation (6) and boundary conditions (7)

$$S(n,r)=S(n-1,r-1)+rS(n-1,r) \qquad (21)$$

where $n, r = 0, 1, 2, 3, \ldots, n \geq r$,

$$S(n,0)=0, n\geq 1,\ S(0,r)=0, r>1, \text{ and } \qquad S(0,0)=1 \qquad (22)$$

What kind of numbers does equation (21) subject to boundary conditions (22) generate? Figure 5.16 shows the spreadsheet that generates those numbers[5]. They have interesting

[5] Numbers $S(n, r)$ are called Stirling numbers of the second kind after James Stirling—a Scottish mathematician of the 18th century. There are also Stirling numbers of the first kind, which are not considered in this book. Thus, in what follows, the specification "second kind" will be omitted for the sake of brevity.

combinatorial meanings representing the number of ways n different objects can be partitioned in r indistinguishable groups. For example, consider

Student Group Problem 1. How many ways can the three students—Al, Beth, and Carol— form two groups?

Solution. It is not difficult to find that there are three ways to form two groups:

$\{AL, BETH\}$ and $\{CAROL\}$, $\{AL, CAROL\}$ and $\{BETH\}$, $\{AL\}$ and $\{BETH, CAROL\}$.

If the notation $S(n, r)$ represents the number of ways r groups can be created out of n students, then $S(3, 2) = 3$ and the spreadsheet of Figure 5.16 (cell *E4*) confirms this result.

Let us solve a slightly more difficult problem.

Student Group Problem 2. How many ways can four students—Al, Beth, Carol, and Don—form two groups?

	A	B	C	D	E	F	G	H	I	J
1	$r\backslash n$	0	1	2	3	4	5	6	7	8
2	0	1	0	0	0	0	0	0	0	0
3	1	0	1	1	1	1	1	1	1	1
4	2	0	0	1	3	7	15	31	63	127
5	3	0	0	0	1	6	25	90	301	966
6	4	0	0	0	0	1	10	65	350	1701
7	5	0	0	0	0	0	1	15	140	1050
8	6	0	0	0	0	0	0	1	21	266
9	7	0	0	0	0	0	0	0	1	28
10	8	0	0	0	0	0	0	0	0	1

Figure 5.16. Stirling numbers of the second kind.

Solution. Note that Don would be added to the situation after Al, Beth, and Carol have formed two groups. For each pair of groups formed by the three students, there are two ways Don can be added:

$\{AL, BETH, DON\}$ and $\{CAROL\}$, or $\{AL, BETH\}$ and $\{CAROL, DON\}$

$\{AL, CAROL, DON\}$ and $\{BETH\}$, or $\{AL, CAROL\}$ and $\{BETH, DON\}$

$\{AL, DON\}$ and $\{BETH, CAROL\}$, or $\{AL\}$ and $\{BETH, CAROL, DON\}$.

The number of groups that can be formed in this way by four students can be described as $2 \cdot S(3,2) = 6$. Yet, there is another possibility that could be overlooked—when Don forms a separate group—that is,

$\{DON\}$ and $\{AL, BETH, CAROL\}$.

The last situation can be described as $S(3,1) = 1$. By the rule of sum,

$$S(4,2) = S(3,1) + 2 \cdot S(3,2).$$

Finally, one can pose (and solve)

Student Group Problem 3. How many ways can n students form r groups?

Solution. First, $n - 1$ students can be arranged in $r - 1$ groups so that the r-th group would always consist of the n-th student not previously present among the $n - 1$ students. This kind of arrangement can be done in $S(n - 1, r - 1)$ ways. In addition, these $n - 1$ students can be arranged in r groups so that the n-th student has r choices to join any of the groups. That can be done in $rS(n - 1, r)$ ways. Noting that, by definition, n students can form r groups in $S(n, r)$ ways, the rule of sum yields equation (21).

In that way, one can begin with modifying equation (6) to the form of equation (21), then model the latter equation within a spreadsheet, and finally, introduce a problem the solutions of which can be found among the numbers generated by the spreadsheet. Then special cases can be extended to interpret the new equation in general terms.

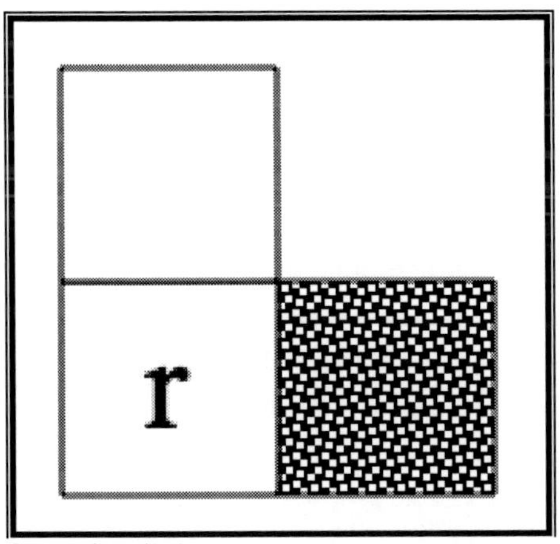

Figure 5.17. Visual imagery of equation (21).

Just like combinations, Stirling numbers can be geometrized by using the results of spreadsheet modeling (Figure 5.16). To this end, any three cells forming an L-shape can be connected as follows: the content of the bottom-right cell is equal to the content of the top-right cell plus the product of the bottom-right cell multiplied by its positional rank in the corresponding column of the spreadsheet template. For example, consider the triple of cells *H6, G6*, and *G5* in the spreadsheet of Figure 5.16. The positional rank of cell *G6* in column *G* is equal to four and

$$65 = 25 + 4 \cdot 10.$$

This relationship is reflected in the diagram of Figure 5.17 that represents the "graph" of Stirling numbers defined by equation (21).

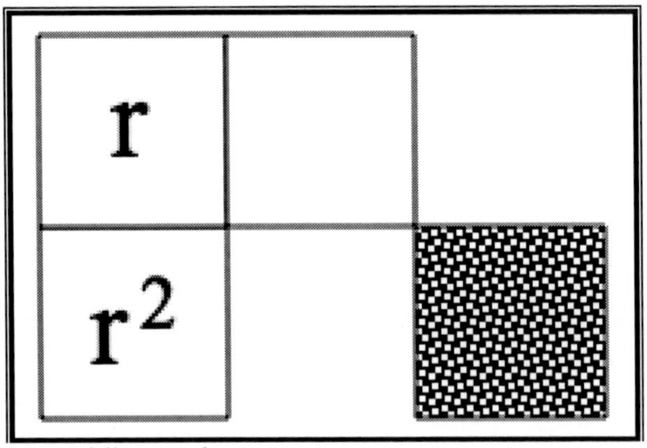

Figure 5.18. Identity among Stirling numbers.

The "graph" of Stirling numbers shown in Figure 5.17 can be transformed to the form shown in Figure 5.18. In turn, the diagram of Figure 5.18 represents the following identity

$$S(n,r)=S(n-1,r-1)+rS(n-2,r-1)+r^2S(n-2,r) \tag{23}$$

Identity (23) can be proved algebraically if one replaces $S(n, r)$ using formula (21). Then,

$$S(n-1,r-1)+rS(n-1,r)=S(n-1,r-1)+rS(n-2,r-1)+r^2S(n-2,r)$$

whence

$$S(n-1,r)=S(n-2,r-1)+rS(n-2,r)=S(n-1,r).$$

Observing Figure 5.16, one can note that the diagonally arranged numbers located immediately above the main (top left-bottom right) diagonal (running from cell *B2* to cell *J10*) are triangular numbers. That is, every Stirling number of the form $S(n, n-1)$ is the triangular number of rank $n-1$. This connection is a special case of a more general identity among the numbers

$$\begin{aligned} S(n,r)&=rS(n-1,r)+(r-1)S(n-2,r-1)\\ &\quad+\ldots+2S(n-r+1,2)+S(n-r,1) \end{aligned} \tag{24}$$

In particular, when $r=n-1$ it follows from (24) that

$$\begin{aligned} &S(n,n-1)=(n-1)S(n-1,n-1)+(n-2)S(n-2,n-2)+\ldots\\ &+2S(n-(n-2),2)+S(n-(n-1),1)=S(1,1)+2S(2,2)+\ldots\\ &+(n-2)S(n-2,n-2)+(n-1)S(n-1,n-1)=1+2+\ldots+(n-2)+(n-1) \end{aligned}$$

due to the fact that $S(k, k) = 1$—meaning that k students can form k groups in one way only. In other words,

$$S(n,n-1)=1+2+3+...+(n-1) \quad (25)$$

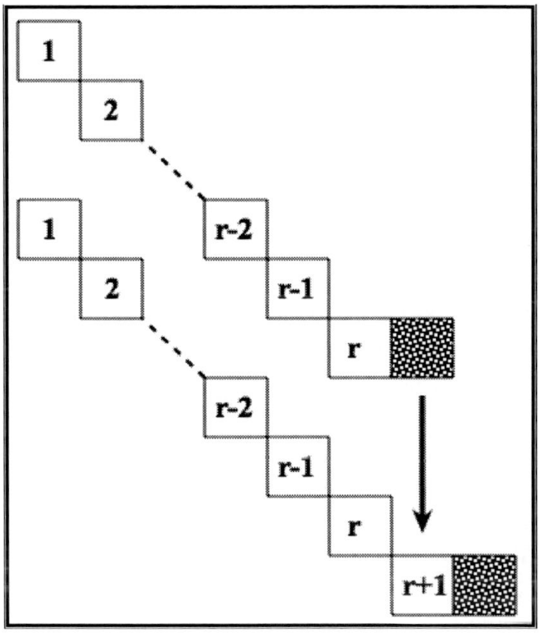

Figure 5.19. Inductive transfer in the case of identity (24).

Formula (25) can be given the following interpretation. The value of $S(n, n - 1)$ represents the number of ways n students can be arranged in $(n - 1)$ groups. Let us numerate the students using consecutive counting numbers, 1 through n. If the first student does not take part in the formation of groups, then the remaining $n - 1$ students can form $n - 1$ groups in one way only. After that, the first student can join any of the groups in $n - 1$ ways. If the first and the second students do not take part in the formation of groups, then the remaining $n - 2$ students can form $n - 2$ groups in one way only. Finally, if only the n-th student takes part in the formation of groups, than a single group can be formed and $n - 1$ students can join the group in one way only. This line of reasoning yields formula (25).

Now, one can prove formula (24) by the method of mathematical induction. The base clause ($r = 1$) means $S(n,1)=S(n-1,1)$—obviously, n students as well as $n - 1$ students can form one group in one way only. Assuming that (24) is true, the transition from (n, r) to $(n + 1, r + 1)$, demonstrated graphically in the diagram of Figure 5.19, can be carried out symbolically as follows:

$$S(n+1,r+1)=S(n,r)+(r+1)S(n,r+1)$$
$$=(r+1)S(n,r+1)+rS(n-1,r)+(r-1)S(n-2,r-2)$$
$$+...+2S(n-r+1,2)+S(n-r,1).$$

This completes the mathematical induction proof of identity (24).

10. Modeling the Sums of Perfect Powers

The sum of the first n counting numbers, $1 + 2 + 3 + \ldots + n$, mentioned in Chapter 1 of this book as the simplest case of a sum of an arithmetic sequence and referred to as the n-th triangular number t_n, represents the simplest example of a sum of perfect powers. Contextually, the number of different rectangles on the $n \times n$ checkerboard is equal to the square of this sum. The goal of this section is to introduce a context for the general case,

$$f(n, r) = 1^r + 2^r + 3^r + \ldots + n^r,$$

where r is a positive integer[6]. To this end, we begin by posing

Phone Number Problem. There are ten digits, 0 through 9, on the touch-pad of a telephone and a phone number consists of four digits to be dialed one by one. How many different phone numbers can be assigned to this telephone?

This problem involves a new type of arrangements. The first digit can be chosen in ten different ways and to each choice of the first digit correspond ten choices of the second digit; to each choice of the first *and* second digits correspond ten choices of the third digit; to each choice of the first, second *and* third digits correspond ten choices of the fourth digit. Therefore, according to the rule of product introduced in section 2 of this chapter, a four-digit phone number can be chosen in 10^4 ways.

The phone number problem can be extended to a more general situation as follows. There are n types of objects and an unlimited number of the objects of each type. One makes up all possible arrangements of r such objects (r-samples) so that each place in an r-sample can be filled in n different ways. Two arrangements are regarded as different if their elements are differently ordered. Arrangements of this type are called r-samples with repetitions of elements of n types.

	A	B	C	D	E	F	G	H
1	$r \backslash n$	1	2	3	4	5	6	7
2	1	1	3	6	10	15	21	28
3	2	1	5	14	30	55	91	140
4	3	1	9	36	100	225	441	784
5	4	1	17	98	354	979	2275	4676
6	5	1	33	276	1300	4425	12201	29008
7	6	1	65	794	4890	20515	67171	184820
8	7	1	129	2316	18700	96825	376761	1200304

Figure 5.20. Sums of perfect powers.

[6] Surprisingly, the case $r = 3$ is equivalent to $(1 + 2 + \ldots + n)^2$, i.e., $f(n,3) = [f(n,1)]^2$ (see also Chapter 1).

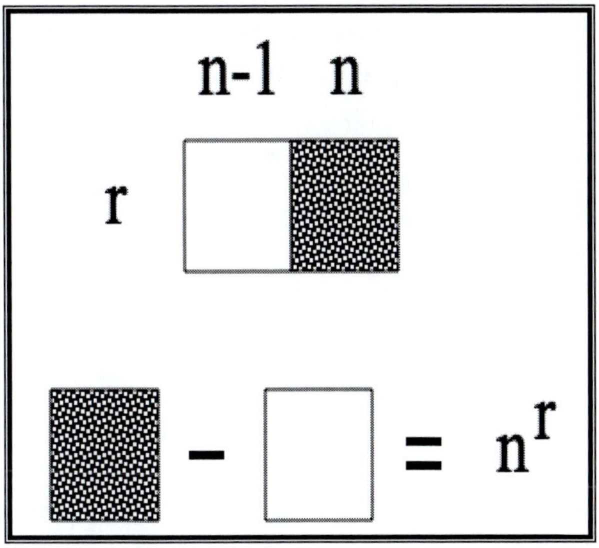

Figure 5.21. The "graph" of equation (26).

By the rule of product, the number of r-samples with repetitions of n objects is equal to n^r. Computing the number of all possible r-samples with repetitions of elements of types varying in the range 1 though n leads to the sum of the r-th powers of the first n counting numbers. This sum is the function of n and r that satisfies the difference equation

$$f(n, r) = f(n-1, r) + n^r \tag{26}$$

subject to the boundary condition in the variable

$$f(1, r) = 1, r \geq 1 \tag{27}$$

Figure 5.20 shows the spreadsheet that generates solutions to equation (26) subject to boundary condition (27).

The geometrical interpretation of equation (26) can be as follows: the difference between contents of every pair of horizontally adjacent cells, is equal to n^r, where (n, r) are the coordinates of the right cell (Figure 5.21).

The appearance of the new array of numbers filled with the sums of perfect powers can motivate search for new connections among integers with different combinatorial meanings. The power of the numerical approach is that it makes possible many mathematical activities based on exploring numerical patterns formed by modeling data. One such activity could be to look for patterns among the numbers located within the first three rows of the spreadsheet of Figure 5.20. This activity often results in conjecturing closed formulas for lower-order sums of perfect powers. Another activity is to compare these numbers (i.e., the sums of perfect numbers) to combinations. As will be shown below, the last activity can lead to the discovery of integers that describe a special class of permutations on the set of counting numbers.

11. Connecting the Sums of Perfect Powers to Combinations

Compare the spreadsheets pictured in Figure 5.20 and Figure 5.2. The numbers in row *2* of the former (sums of counting numbers) and in row *4* (beginning from cell *D4*) of the latter (two-combinations) coincide. Algebraically, this comparison yields

$$f(n, 1) = C(n+1, 2) \tag{28}$$

Relation (28) can be proved using the method of mathematical induction. To this end, note that when $n = 1$, relation (28) turns into the (true) equality $f(1, 1) = C(2, 2)$. Assuming that (28) is true, one has to show that substituting $n + 1$ for n results in $f(n+1,1) = C(n+2,2)$.

Indeed, using definition (26) and (assumed to be true) relation (28) yields $f(n+1,1) = f(n,1) + n + 1 = C(n+1,2) + (n+1)$
$= C(n+1,2) + C(n+1,1) = C(n+2,2)$

This completes the mathematical induction proof of identity (28).

Another way to prove identity (28) is to count the number of ways of creating two-combinations from among $n + 1$ objects by using combinatorial reasoning. To this end, the first object can be paired with n other objects in n ways, the second object can be paired with the $n - 1$ other objects (not including the first one) in $n - 1$ ways, the third object can be paired with $n - 2$ objects (not including the first and the second ones) in $n - 2$ ways, and so on. Finally, the nth object can be paired with the $(n + 1)^{th}$ object in one way only. Therefore,

$$C(n+1,2) = 1 + 2 + ... + (n-1) + n = f(n,1).$$

As an example of creating two-combinations from among $n + 1$ objects consider one more time the problem of counting rectangles on the $n \times n$ checkerboard briefly mentioned in the beginning of section 10 and discussed in detail in Chapter 1. As the checkerboard includes $n + 1$ horizontal lines, there are $C(n+1,2)$ ways to choose a pair of top and bottom sides of a rectangle. The same is true in the case of choosing its left and right sides. The rule of product (section 2) yields $[C(n+1,2)]^2$—the total number of rectangles on the $n \times n$ checkerboard. Finally, the use of formula (28) confirms formula (8) of Chapter 1.

Now, compare the numbers in rows 3 and 5 of the spreadsheets pictured in Figure 5.20 and Figure 5.2, respectively. Each sum of two consecutive positive integers in row 5 can be found among integers in row 3, for example,

$$f(2,2)=C(3,3)+C(4,3), f(3,2)=C(4,3)+C(5,3), f(4,2)=C(5,3)+C(6,3),$$

and so on. This observation can be generalized as follows

$$f(n,2)=C(n+1,3)+C(n+2,3) \tag{29}$$

Identity (29) can be proved by the method of mathematical induction. When $n = 1$, relation (29) holds true as $1 = C(2, 3) + C(3, 3) = 0 + 1 = 1$. The essence of transition from n to $n + 1$ consists in proving the equality

$$f(n+1, 2) = C(n+2,3) + C(n+3,3)$$

under the assumption that (29) holds true. To this end, one can show that the difference between the left-hand side and the right-hand side of the last equality is equal to zero. The required computations are not difficult and can be carried out without a technological support. Indeed,

$$\begin{aligned}
&f(n+1,2) - C(n+2,3) - C(n+3,3) \\
&= f(n,2) + (n+1)^2 - C(n+2,3) - C(n+3,3) \\
&= (n+1)^2 + C(n+1,3) + C(n+2,3) - C(n+2,3) - C(n+3,3) \\
&= (n+1)^2 + C(n+1,3) - C(n+3,3) \\
&= (n+1)^2 + C(n+2,3) - C(n+1,2) - C(n+2,3) - C(n+2,2) \\
&= (n+1)^2 - 2C(n+1,2) - C(n+1,1) = (n+1)^2 - 2\frac{n(n+1)}{2} + n + 1 \\
&= (n+1)^2 - (n+1)^2 = 0.
\end{aligned}$$

This completes the proof of identity (29).

The next task is to compare numbers in rows 4 and 6 of the spreadsheets pictured in Figures 5.20 and 5.2, respectively. To this end, note that the sum of the first n counting numbers consisted of one combination, the sums of the first n squares of counting numbers consisted of the sum of two combinations and, thereby, one can conjecture that the sum of the first n cubes of counting numbers is equal to the sum of three combinations. A few attempts to find this kind of connection fail, however, when in such sum all coefficients are equal to one. A trial and error approach may bring about the following special cases

$$f(2,3) = C(3,4) + 4C(4,4) + C(5,4),$$

$$f(3,3) = C(4,4) + 4C(5,4) + C(6,4),$$

$$f(4,3) = C(5,4) + 4C(6,4) + C(7,4).$$

Generalizing from the three special cases yields the conjecture

$$f(n,3) = C(n+1,4) + 4C(n+2,4) + C(n+3,4) \qquad (30)$$

The "graph" of combinations shown in Figure 5.3 can guide the mathematical induction proof of this conjecture. When $n = 1$ we have $C(2,4) + 4C(3,4) + C(4,4) = 1$ and $f(1, 3) = 1$. Transition from n to $n + 1$ in proving relation (30) requires the following demonstration

$$(n+1)^3 = f(n+1,3) - f(n,3) = C(n+2,4) + 4C(n+3,4) + C(n+4,4)$$
$$-C(n+1,4) - 4C(n+2,4) - C(n+3,4)$$
$$= -3C(n+2,4) + 3C(n+3,4) + C(n+4,4) - C(n+1,4).$$

This time, the required computations are more involved and the use of *Maple* in the completion of proof of identity (30) is justified. Figure 5.22 shows all steps required to demonstrate that

$$-3C(n+2,4) + 3C(n+3,4) + C(n+4,4) - C(n+1,4) = (n+1)^3.$$

This completes the proof of identity (30).

Continuing in this vein, one has to find the representation of the sum of the first n fourth powers of counting numbers, $f(n,4)$, through combinations. The first plausible conjecture is that there should be four additive terms of the form $C(k, 5)$ in this representation where k varies from n to $n + 4$. The second conjecture pertains to the value of coefficients in those terms. Analyzing the coefficients in the right-hand sides of identities (28)-(30) suggests the development which resembles that of Pascal's triangle; more specifically,

$$f(n,4) = C(n+1,5) + xC(n+2,5) + xC(n+3, 5) + C(n+4,5) \qquad (31)$$

Note that $n = 2$ is the smallest value of n for which at least one term in (31) that contains x does not vanish. Substituting $n = 2$ in (31) yields

$C(n, r) := \dfrac{n!}{r! \cdot (n-r)!}$

$$(n, r) \rightarrow \frac{n!}{r!\,(n-r)!}$$

$A1 := -3 \cdot C(n+2, 4)$

$$-\frac{1}{8}\frac{(n+2)!}{(n-2)!}$$

$A2 := 3 \cdot C(n+3, 4)$

$$\frac{1}{8}\frac{(n+3)!}{(n-1)!}$$

$A3 := C(n+4, 4)$

$$\frac{1}{24}\frac{(n+4)!}{n!}$$

$A4 := -C(n+1, 4)$

$$-\frac{1}{24}\frac{(n+1)!}{(n-3)!}$$

$A5 := A1 + A2 + A3 + A4$

$$-\frac{1}{8}\frac{(n+2)!}{(n-2)!} + \frac{1}{8}\frac{(n+3)!}{(n-1)!} + \frac{1}{24}\frac{(n+4)!}{n!} - \frac{1}{24}\frac{(n+1)!}{(n-3)!}$$

$factor(\%)$

$$(n+1)^3$$

Figure 5.22. Using *Maple* in proving (30) by mathematical induction.

$$17 = 0 + x \cdot 0 + x \cdot 1 + 6$$

whence $x = 11$.

Therefore, relation (31) assumes the form of identity

$$f(n,4) = C(n+1,5) + 11C(n+2,5) + 11C(n+3,5) + C(n+4,5) \qquad (32)$$

Once again, one can use *Maple* in proving identity (32). The proof is based on the demonstration that

$$\begin{aligned}
&(n+1)^4 = f(n+1,4) - f(n,4) \\
&= C(n+2,5) + 11C(n+3,5) + 11C(n+4,5) + C(n+5,5) \\
&-C(n+1,5) - 11C(n+2,5) - 11C(n+3,5) - C(n+4,5) \\
&= -10C(n+2,5) + 10C(n+4,5) - C(n+1,5) + C(n+5,5).
\end{aligned}$$

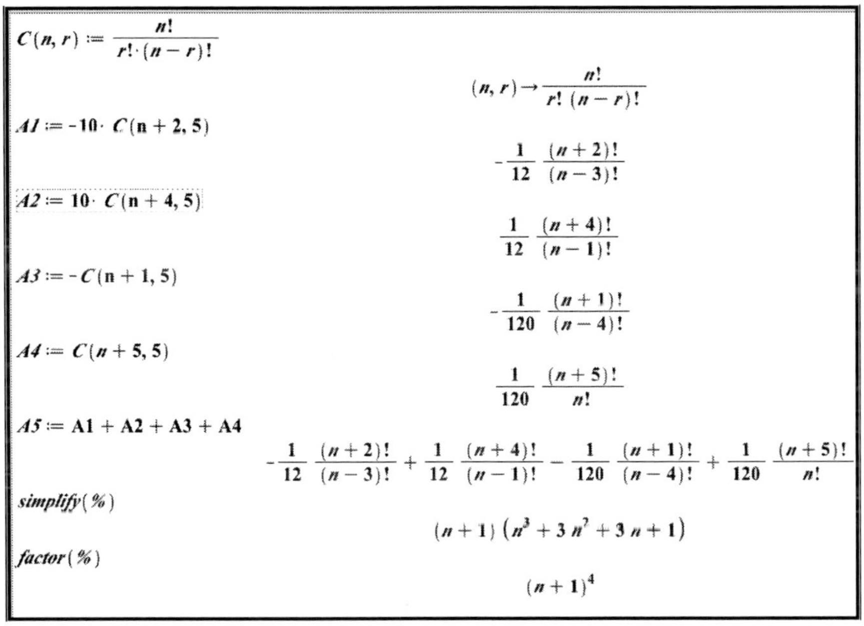

Figure 5.23. *Maple* proof of identity (32).

The necessary steps connecting the far right and the far left parts of the last chain of equalities are shown in Figure 5.23.

In much the same way, one may conjecture that

$$f(n,5) = C(n+1,6)+xC(n+2,6)+yC(n+3,6)+xC(n+4,6)+C(n+5,6) \qquad (33)$$

Now, the goal is to construct a system of two linear equations in variables x and y by using two, preferably the smallest possible, values of n. Note that $n = 3$ is the smallest value that can be plugged into equation (33) in order to turn into zero the first two terms in its right-hand side only. This yields

$$276 = yC(6,6) + xC(7,6) + C(8,6)$$

whence

$$7x + y = 248$$

Next, substituting 4 for n in equation (33) yields

$$1300 = xC(6,6) + yC(7,6) + xC(8,6) + C(9,6)$$

whence

$$29x + 7y = 1216$$

To find the values of x and y satisfying the above two linear equations, one can use the following *Maple* code

solve({7·*x* + *y* = 248, 29·*x* + 7·*y* = 1216}, [*x*, *y*])

yielding $x = 26$ and $y = 66$. Therefore,

$$\begin{aligned} f(n,5) &= C(n+1,6) + 26C(n+2,6) \\ &+66C(n+3,6) + 26C(n+4,6) + C(n+5,6) \end{aligned} \qquad (34)$$

One can see that the coefficients in the right-hand sides of relations (28)-(30), (32), and (34) form the following array

$$\begin{matrix} 1 & 0 & 0 & 0 & 0 \\ 1 & 1 & 0 & 0 & 0 \\ 1 & 4 & 1 & 0 & 0 \\ 1 & 11 & 11 & 1 & 0 \\ 1 & 26 & 66 & 26 & 1 \end{matrix} \qquad (35)$$

resembling Pascal's triangle.

12. EULERIAN NUMBERS

Array (35), the elements of which will be denoted $A(n,r)$, where n is the vertical coordinate and r is the horizontal coordinate, represents a matrix, the first column and the first row of which consist, respectively, of ones and zeros. In order to find a rule that produces the numbers $A(n,r)$, consider the 5th row of the array ($n = 5$). A possible guess is that the number

11 and the number 1 from the 4throw both contribute to the number 26 in the 5th row, whilst, in turn, the number 11 arises as a combination of the numbers 1 and 4 from the 3rd row, etc. The goal is to develop a partial difference equation that describes this recursive process. A pure observation brings about a few special cases

$A(3,2) = 2A(2, 1) + 2A(2,2)$, that is, $4 = 2\cdot 1 + 2\cdot 1$,

$A(4,2) = 3A(3, 1) + 2A(3, 2)$, that is, $11 = 3\cdot 1 + 2\cdot 4$,

$A(5,3) = 3A(4, 2) + 3A(4, 3)$, that is, $66 = 3\cdot 11 + 3\cdot 11$.

These relations resemble a non-linear form of the recurrence that define Stirling numbers. In other words, a plausible guess is that the coefficients in $A(n, r)$ in the above three numerical recurrences could be the functions of n and r.

In order to discover a general rule that generates these coefficients, note that in all the three cases the second coefficient coincides with the second coordinate of $A(n, r)$ located on the left. Next, consider each first coefficient as a linear function $L(n, r) = xn + yr + z$ of the both coordinates, n and r. This assumption results in following system of equations

$$3x + 2y + z = 2$$

$$4x + 2y + z = 3$$

$$5x + 3y + z = 3$$

Which solution $[x = 1; y = -1; z = 1]$ can be found by using the following *Maple* code:

```
solve({3·x + 2·y + z = 2, 4·x + 2·y + z = 3, 5·x + 3·y + z = 3}, [x, y, z]).
```

The triple (1, -1, 1) turns $L(n, r) = xn + yr + z$ into the sum $n - r + 1$. This, together with an earlier guess about the second coefficient being equal to n, allows one to conjecture the recurrence (partial difference) equation

$$A(n, r) = (n - r + 1)A(n - 1, r - 1) + rA(n, r - 1) \quad (36)$$

which, in particular, generates numbers shown in array (35). Boundary conditions for difference equation (36), derived from this array, can be described in the following form

$$A(n, 1) = 1, n \geq 1; A(1, r) = 0, r > 1 \quad (37)$$

The numbers $A(n, r)$ generated by equation (36) subject to boundary conditions (37) are known as Eulerian numbers (not to be confused with Euler numbers) introduced by Euler in 1755 as the coefficients of certain polynomials (that also bear his name).

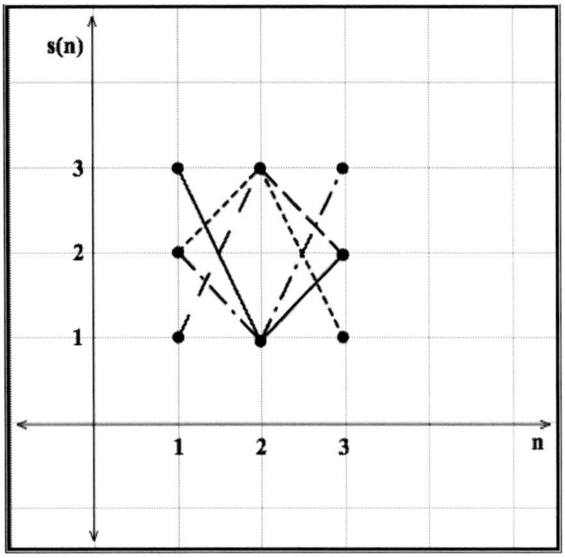

Figure 5.24. Permutations on Z_3 with two rises.

Eulerian numbers have an interesting combinatorial interpretation. Let

$$s = [s(1), s(2), ..., s(r)]$$

be a permutation on the set of the first n counting numbers $Z_n = \{1, 2, ..., n\}$. It is said that the permutation s has exactly r rises if there exist exactly $r - 1$ values of i such that the inequality $s(i) < s(i+1)$ holds. Then the total number of permutations on Z_n that have r rises is precisely the number $A(n, r)$. One can verify this interpretation for a few first lines of array (35). For example, permutation [3, 1, 2] on the set $Z_3 = \{1, 2, 3\}$ has two rises because only the second and the third elements are such that $s(2) = 1 < 2 = s(3) = 2$. The other three permutations on Z_3 with two rises are[1, 3, 2], [2, 3, 1], and [2, 1, 3]. Therefore the total number of such permutations (presented graphically in Figure 5.24 where each of the two-sided polygonal lines has only one side with a positive slope) is equal to four—the value of $A(3, 2)$ in array (35).

Recall (section 2) that the total number of permutations on Z_n equals $n!$ When $n = 3$ we have $3! = 6$—the sum of Eulerian numbers in row 3 of array (35). This observation is consistent with the fact that [3, 2, 1] and [1, 2, 3] are the only permutations on Z_3 with one and three rises, respectively. In other words, each permutation on Z_3 has either one, or two, or three rises. Furthermore, the sums of Eulerian numbers in rows 4 and 5 are equal to 4! (the number of permutations on Z_4) and 5! (the number of permutations on Z_5), respectively.

In particular, one such permutation on the set Z_5 is [1, 5, 3, 4, 2] and it has exactly three rises (indeed, only $1 < 5$ and $3 < 4$) like 65 others included in $A(5, 3)$. In general, each permutation on Z_n has either 1, or 2, …, or n rises. Now, the fact that any row in array (35) starts and ends with the number 1 has a simple combinatorial interpretation. These boundary elements of the array represent, respectively, the number of permutations with the smallest and the largest number of rises, for each value of n. So, the only permutations on Z_5 with one and five rises are [5, 4, 3, 2, 1] (a descending permutation) and [1, 2, 3, 4, 5] (an ascending permutation), respectively.

	A	B	C	D	E	F	G	H	I	J	K
1	n\r	1	2	3	4	5	6	7	8	9	10
2	1	1	0	0	0	0	0	0	0	0	0
3	2	1	1	0	0	0	0	0	0	0	0
4	3	1	4	1	0	0	0	0	0	0	0
5	4	1	11	11	1	0	0	0	0	0	0
6	5	1	26	66	26	1	0	0	0	0	0
7	6	1	57	302	302	57	1	0	0	0	0
8	7	1	120	1191	2416	1191	120	1	0	0	0
9	8	1	247	4293	15619	15619	4293	247	1	0	0
10	9	1	502	14608	88234	156190	88234	14608	502	1	0
11	10	1	1013	47840	455192	1310354	1310354	455192	47840	1013	1

Figure 5.25. Eulerian numbers generated by a spreadsheet.

One can use a spreadsheet to generate Eulerian numbers through modeling equation (36) subject to conditions (37). The spreadsheet filled with numbers $A(n,r)$ is shown in Figure 5.25. One can also check to see that the sum of numbers in each row of the spreadsheet is the factorial of the row number. That is,

$$n! = A(n,1) + A(n,2) + ... + A(n,n)$$

or

$$\sum_{i=1}^{n} A(n,i) = \prod_{i=1}^{n} i .$$

The last relation gives yet another example of the intrinsic connectivity of mathematical concepts.

Generalizing from identities (28)-(30), (32), and (34) one can arrive at the formula

$$f(n,r) = A(r,1)C(n+1,r+1) + A(r,2)C(n+2,r+1) +. .. + A(r,r)C(n+r,r+1)$$

which can be rewritten in the form

$$1^r + 2^r + 3^r + ... + n^r = \sum_{i=1}^{r} A(r,i)C(n+i,r+1) \qquad (38)$$

It follows from formula (38) that

$$n^r = f(n,r) - f(n-1,r)$$
$$= \sum_{i=1}^{r} A(r,i)C(n+i,r+1) - \sum_{i=1}^{r} A(r,i)C(n-1+i,r+1)$$
$$= \sum_{i=1}^{r} A(r,i)[C(n+i,r+1) - C(n-1+i,r+1)] = \sum_{i=1}^{r} A(r,i)C(n-1+i,r).$$

Here formula (6) was used to replace the difference of two combinations by a single combination. In that way,

$$n^r = \sum_{i=1}^{r} A(r,i)C(n-1+i,r) \tag{39}$$

Formula (39) is known as Worpitzky's formula[7]. In particular, this formula, by representing repeated multiplication as a combination of integers with combinatorial meanings, demonstrates one of the most profound features of mathematics—a possibility of representing simple structures through more and more complex ones. For example, in the case $n = 10$ and $r = 2$ formula (39) yields

$$10^2 = A(2,1)\cdot C(10,2) + A(2,2)\cdot C(11,2) = 1\cdot 45 + 1\cdot 55 = 100.$$

In other words, the number 10 multiplied by itself is equal to the product of the number of permutations on the set {1, 2} with one rise and the number of two-combinations of ten objects plus the number of permutations on {1, 2} with two rises and the number of two-combinations of eleven objects.

Other representations of the power 10^2 include:

$10^2 = 1+3+5+7+9+11+13+15+17+19$—the sum of the first ten odd numbers;

$10^2 = 45+55$—the sum of the 9th and 10th triangular numbers, or, alternatively, the sum of two-combinations of ten objects and eleven objects;

$10^2 = 8^2 + 6^2$—the sum of two squares or the Pythagorean triple (10, 8, 6);

$10^2 = 1^3 + 2^3 + 3^3 + 4^3$—the sum of the first four cubes of counting numbers.

As the National Council of Teachers of Mathematics (2000) put it: "When students can see the connections across content areas, they develop a view of mathematics as a integrated whole" (p. 355). This explains why teachers' leaning to recognize and appreciate mathematical connections should be continuously emphasized in a capstone course.

---

[7] Julius Worpitzky—a German mathematician of the 19th century who published deep mathematical results in annual reports of secondary schools while working as a teacher at Friedrich-Gymnasium in Berlin. It was not uncommon for German mathematicians of that time to teach secondary school for a number of years. For example, both Kummer and Weierstrass were teachers and published some of their results in secondary school reports (Jacobsen, Thron, and Waadeland, 1989).

13. Closed Formulas for the Sums of Perfect Powers

One can use formula (38), the spreadsheet of Figure 5.25, and *Maple* (in calculating combinations and simplifying algebraic expressions) to obtain closed formulas for the sums of the first n perfect powers.

1) When $r = 1$ we have (cf. formula (2), Chapter 1)

$$1+2+3+...+n=\sum_{i=1}^{1} A(i,1)\cdot C(n+i,2) = A(1,1)\cdot C(n+1,2) = \frac{n(n+1)}{2}.$$

That is,

$$1+2+3+...+n=\frac{n(n+1)}{2}.$$

2) When $r = 2$ we have (cf. formula (4), Chapter 1)

$$1^2+2^2+3^2+...+n^2=\sum_{i=1}^{2} A(i,2)C(n+i,3) = C(n+1,3)+C(n+2,3)$$

$$=\frac{n(n+1)(2n+1)}{6}.$$

That is,

$$1^2+2^2+3^2+...+n^2=\frac{n(n+1)(2n+1)}{6}.$$

3) When $r = 3$ we have (cf. formula (5), Chapter 1)

$$1^3+2^3+3^3+...+n^3$$

$$=\sum_{i=1}^{3} A(i,3)C(n+i,4) = C(n+1,4)+4C(n+2,4)+C(n+3,4)$$

$$=\frac{n^2(n+1)^2}{4}.$$

That is,

$$1^3+2^3+3^3+...+n^3=\frac{n^2(n+1)^2}{4}.$$

4) When $r = 4$ we have

$$1^4 + 2^4 + 3^4 + \ldots + n^4 = \sum_{i=1}^{4} A(i,4)C(n+i,5)$$
$$= C(n+1,5) + 11C(n+2,5) + 11C(n+3,5) + C(n+4,5)$$
$$= \frac{n(n+1)(2n+1)(3n^2+3n-1)}{30}.$$

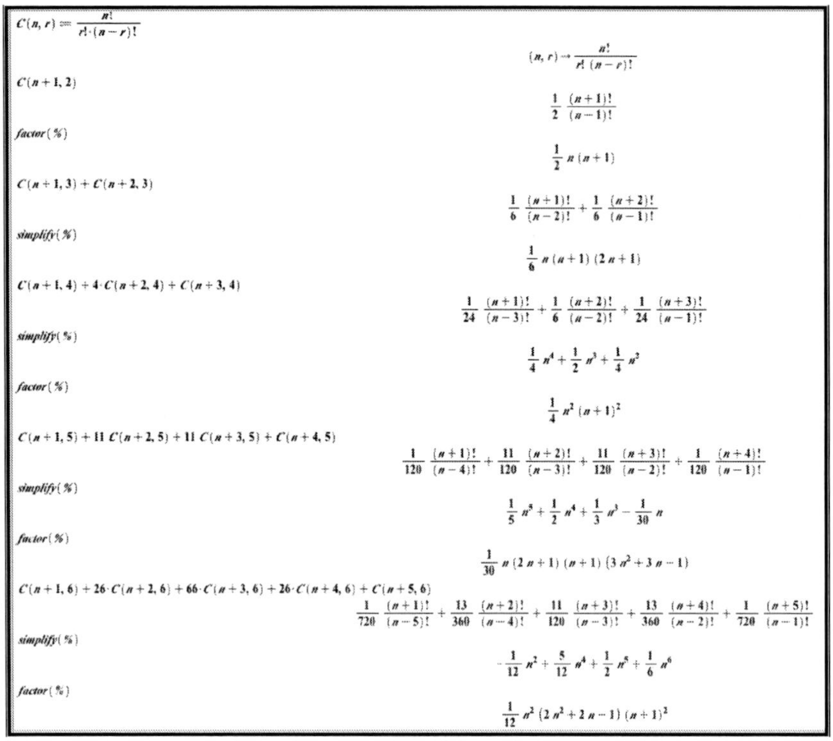

Figure 5.26. Using *Maple* in symbolic computations of formula (38).

That is,

$$1^4 + 2^4 + 3^4 + \ldots + n^4 = \frac{n(n+1)(2n+1)(3n^2+3n-1)}{30}.$$

5) When $r = 5$ one can add fifth powers of integers as follows:

$$1^5 + 2^5 + 3^5 + \ldots + n^5 = \sum_{i=1}^{5} A(i,5)C(n+i,6)$$
$$= C(n+1,6) + 26C(n+2,6) + 66C(n+3,6) + 26C(n+4,6) + (n+5,6)$$
$$= \frac{n^2(n+1)^2(2n^2+2n-1)}{12}.$$

That is,

$$1^5 + 2^5 + 3^5 + ... + n^5 = \frac{n^2(n+1)^2(2n^2+2n-1)}{12}.$$

Figure 5.26 shows *Maple*-supported symbolic computations in the development of the above five closed formulas for the sums of the first *n* perfect powers.

14. Activity Set

1. Rewrite identities for $\bar{C}(n,r)$ in terms of $C(n,r)$ and prove new identities.
2. Prove that Stirling numbers of the second kind satisfy the identity $S(n,r) = rS(n-1,r) + (r-1)S(n-2,r-1) + S(n-2,r-2)$ and draw its "graph."
3. How many ways can Anna put six rings on her fingers, excluding the thumbs?
4. Three segments are given whose lengths are 3, 4, and 5 inches. Using any of the given lengths as many times as you wish, determine how many equilateral triangles can be constructed.
5. How many ways can five students form three groups?
6. Show that the number of ways five students can form three groups is equal to the number of ways these five students can form two groups and four groups. Can this situation be extended to six students forming three, two, and four groups, respectively?
7. Using *Maple* show that
   $f(n,6) = C(n+1,7) + 57C(n+2,7) + 302C(n+3,7)$
   $+302C(n+4,7) + 57C(n+5,7) + C(n+6,7)$.
8. How many permutations of the elements of the set {1, 2, 3, 4, 5, 6} are there? How many permutations with exactly one rise are there on this set? How many permutations with exactly six rises are there on this set?

Chapter 6

HISTORICAL PERSPECTIVES

> Thus, according to Plato, it is in mathematics that one learns to reason concerning things in themselves, that one receives training in dialectics. Mathematical objects lie, in his view, between the visible things and the ideas, and mathematical understanding stands between opinion and philosophical insight. These are the reasons why Plato attributed such eminent importance to mathematical training.
>
> — Van der Waerden (1961, p.149)

1. INTRODUCTION

The use of historical examples as a background for teaching mathematics to teachers has long been recognized as a useful didactical tool. It has been argued (Conference Board of the Mathematical Sciences, 2001) that appropriately selected historical contexts for a capstone course enhanced by the use of modern tools of technology can provide teachers with "insight for teaching that they are unlikely to acquire in courses for mathematics majors headed to graduate school or technical work" (p. 127). Towards this end, a number of investigations, concepts, and geometric constructions known from the history of mathematics that span from antiquity to the 19th century will be introduced in this chapter.

One geometric construction can be traced back to Theodorus of Cyrene (a city on the north coast of Africa)—a Pythagorean philosopher of the 5th century B. C., known to be a teacher of Plato. In particular, Theodorus is credited for constructing the spiral pictured in Figure 6.1 to support his studies of line segments, incommensurable in length. Sides of the squares with areas 3 and 5 square units are examples of the segments of that kind.

In connection with the spiral of Theodorus, another historical context will be discussed in this chapter. It concerns the construction of square—the main geometric figure in the Sulvasutras. This ancient—between the 6th and 3rd centuries B.C. (Thibaut, 1984)—Indian geometry text, known also as "rules of the cord," refers to the cord (line segment) as a generator of area. A typical theorem from the Sulvasutras reads: "The cord stretched in the diagonal of an oblong produces both [originally given areas] which the cords forming the longer and shorter sides produce" (Coolidge, 1963, p. 14). In this formulation, the main object is the cord, a producer of area of the square constructed on the diagonal of a rectangle. As the spiral of Theodorus consists of right triangles, their sides can be construed as the producers of area.

Whereas all the squares constructed on the diagonal of rectangles are similar geometrically, their areas possess different numerical properties. Indeed, following the "atomistic" idea used by the early Pythagoreans to represent numbers as geometric patterns, counting numbers can be associated with regular polygons and, thereby, can be construed as polygonal numbers[1]. Thus the elements of the right triangles in the spiral of Theodorus can be expressed as the square roots of the polygonal numbers. For example, the triangular numbers 1, 3, 6, 10, 15 represent areas produced by the legs of the right triangles which positional rank within the spiral (counting in the clock-wise order starting from the smallest triangle) are equal to these numbers, respectively. The same can be said about the square numbers 1, 4, 9, 16; the pentagonal numbers 1, 5, 12; the hexagonal numbers 1, 6, 15 (to list only numbers appearing in the spiral of Theodorus). In general, any counting number $m \geq 3$ can be construed as a polygonal number of side m and rank two.

Below, by integrating various historical contexts, the elements of the spiral of Theodorus (and of its extension) will be used to discover and then prove various relationships among areas expressed through polygonal numbers. It will also be shown how one can use formal proof as a context for posing (and solving) new problems, thereby enhancing one's mathematical content knowledge. In that way, it will be shown how the spiral of Theodorus, in addition to enabling research in pure mathematics (Davis, 1993), can motivate the development of new environments for mathematics education, in general, and for a capstone course, in particular.

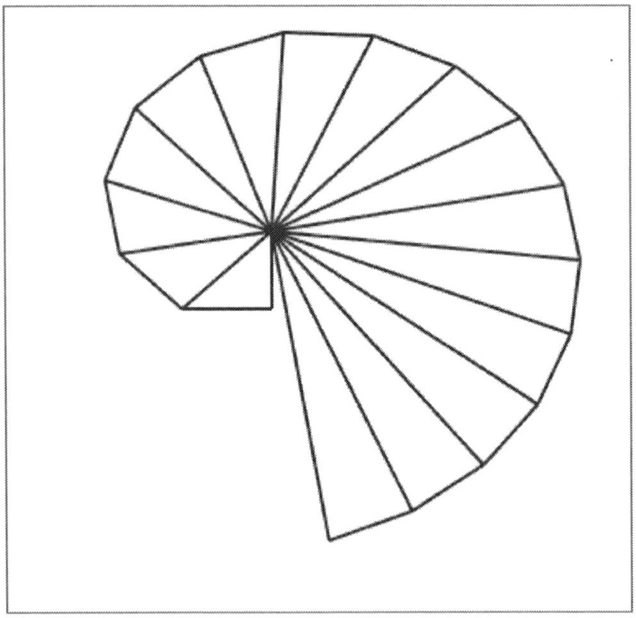

Figure 6.1. The spiral of Theodorus.

[1] Triangular numbers introduced in Chapter 1 represent the simplest example of polygonal numbers. Square and pentagonal numbers were briefly mentioned in Chapter 4.

2. The Spiral of Theodorus Motivates Concept Learning with Technology

The original spiral of Theodorus was limited to 16 triangles (Figure 6.1). One can see that the construction of the 17th right triangle (having the hypotenuse $\sqrt{18}$) would have resulted in overlapping triangles, thereby making a figure rather inconvenient for further analysis. Going beyond 16 triangles and using the spiral built on the multiple application of the Pythagorean theorem, allows for a geometric demonstration of how the sequence of the square roots of consecutive counting numbers

$$f_n = \sqrt{n}\,,\ \ n = 1, 2, 3, \ldots \tag{1}$$

can be defined recursively in the form of a non-linear difference equation of the first order

$$f_n = \sqrt{f_{n-1}^{\ 2} + 1}, \quad f_1 = 1 \tag{2}$$

n	$n^{0.5}$	f_n
1	1	1
2	1.4142	1.4142
3	1.7321	1.7321
4	2	2
5	2.2361	2.2361
6	2.4495	2.4495
7	2.6458	2.6458
8	2.8284	2.8284
9	3	3
10	3.1623	3.1623
11	3.3166	3.3166
12	3.4641	3.4641
13	3.6056	3.6056
14	3.7417	3.7417
15	3.873	3.873
16	4	4
17	4.1231	4.1231

Figure 6.2. Generating square roots through closed and recursive definitions.

Relations (1) and (2) represent, respectively, closed and recursive definitions of the square root of a counting number n. By using a spreadsheet as an iterative tool, one can see the equivalence of the two definitions in a numeric form (Figure 6.2). Whereas the meaning of definition (1) is easy to understand just by its appearance, the square roots of consecutive counting numbers are not explicitly present in definition (2) and the spiral of Theodorus can be used to reveal the hidden meaning of this non-linear recurrence relation.

In what follows, a number of interesting properties of the spiral will be discussed, and connections between different analytical representations of other mathematical concepts will be established. Through this discussion it will be demonstrated how the extension of the original spiral encouraged and enhanced by the use of technology brings about a number of geometric and analytic discoveries associated with its metric properties. Apparently, such discoveries were not possible in the ancient environment of the stick and sand available to Theodorus.

3. The Construction of the Spiral of Theodorus by Using a Spreadsheet

The very construction of the spiral of Theodorus can be done using spreadsheet's facility to generate geometric figures from numerical tables. Such use of a spreadsheet integrates the ancient tradition of the geometrization of mathematical structures and modern pedagogical position that the use of technology enhances concept learning. In order to derive formulas needed for this construction, consider the spiral in the coordinate plane (x, y). By definition, the nth triangle of the spiral (counting in the clock-wise order) has the side lengths $1, \sqrt{n}$ and $\sqrt{n+1}$, $n \geq 1$. Let us assume that the center of the spiral resides at the origin and an endpoint of the (outer) unit leg of the nth triangle has the coordinates (x_n, y_n), $n = 1, 2, 3, \ldots$. Applying the distance formula to the pairs of points $\{(x_n, y_n), (x_{n-1}, y_{n-1})\}$ and $\{(x_n, y_n), (0,0)\}$ (see Figure 6.3) yields the system of simultaneous equations

$$x_n^2 + y_n^2 = n+1 \tag{3}$$

$$(x_n - x_{n-1})^2 + (y_n - y_{n-1})^2 = 1 \tag{4}$$

Proposition 1. The system of equations (3), (4) has the following iterative solution

$$x_n = x_{n-1} + \frac{y_{n-1}}{\sqrt{n}}, \quad y_n = y_{n-1} - \frac{x_{n-1}}{\sqrt{n}}, \quad x_0 = 0, \quad y_0 = -1 \tag{5}$$

Proof. In order to derive iterative formulas for the coordinates x_n and y_n, the left-hand side of equation (4) can be expanded to the form

$$(x_n^2 + y_n^2) + (x_{n-1}^2 + y_{n-1}^2) - 2(x_n x_{n-1} + y_n y_{n-1}) = 1 \tag{6}$$

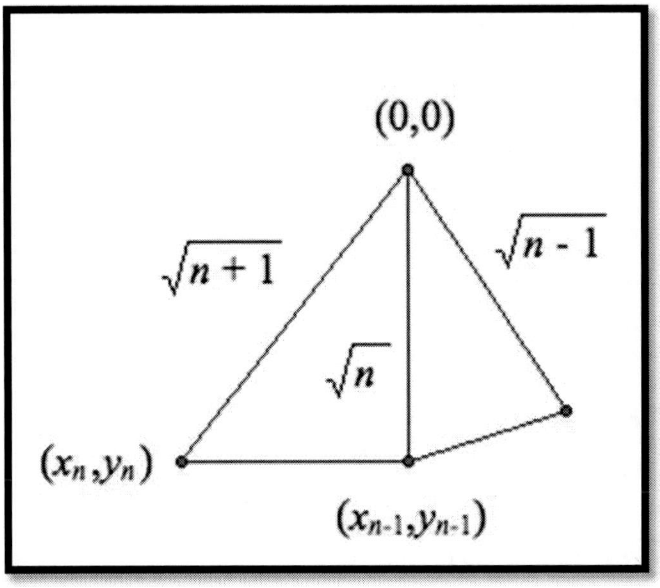

Figure 6.3. Developing equations (3) and (4).

Using jointly (6) and (3) yields

$$n+1+n-2(x_n x_{n-1}+y_n y_{n-1})=1$$

from where it follows that

$$x_n x_{n-1}+y_n y_{n-1}=n$$

whence

$$y_n=\frac{n-x_n x_{n-1}}{y_{n-1}} \tag{7}$$

Now, using jointly (7) and (3) yields the equation

$$x_n^2+(\frac{n-x_n x_{n-1}}{y_{n-1}})^2=n+1$$

which can be transformed to a quadratic equation in respect to x_n as follows

$$n^2+x_n^2 x_{n-1}^2-2nx_n x_{n-1}+y_{n-1}^2 x_n^2=y_{n-1}^2(n+1)$$

or

$$(x_{n-1}^2+y_{n-1}^2)x_n^2-2nx_{n-1}x_n+n^2-y_{n-1}^2(n+1)=0$$

Applying relation (3) to the coefficient in x_n^2 yields the equation

$$nx_n^2-2nx_{n-1}x_n+n^2-y_{n-1}^2(n+1)=0 \tag{8}$$

As the center of the spiral belongs to the origin, one has $(x_0,y_0)=(0,-1)$ and, in order for x_1 to be smaller than zero, the sequence x_n should be defined as the following root of equation (8)

$$x_n=\frac{nx_{n-1}+\sqrt{n^2x_{n-1}^2-n(n^2-ny_{n-1}^2-y_{n-1}^2)}}{n} \tag{9}$$

Once again, using equation (3) to simplify the radicand in (9) yields

$$x_n=\frac{nx_{n-1}+\sqrt{n^2(x_{n-1}^2+y_{n-1}^2)-n^3+ny_{n-1}^2)}}{n}=\frac{ny_{n-1}+\sqrt{n^3-n^3+ny_{n-1}^2}}{n}$$

$$=\frac{nx_{n-1}+\sqrt{n}y_{n-1}}{n}=x_{n-1}+\frac{y_{n-1}}{\sqrt{n}}.$$

Therefore,

$$x_n=x_{n-1}+\frac{y_{n-1}}{\sqrt{n}} \tag{10}$$

Note that the plus sign in the right-hand side of formula (10) is due to the fact that the transition from the point (x_0, y_0) = (0, -1) to the point (-1, y_1) and $x_1=0+\frac{-1}{1}=-1$. Had we taken the square root in formula (9) with the minus sign, we would have had x_1 = 1 and, thereby, the spiral would have developed in an opposite direction.

Now, using jointly (10) and (7) yields

$$y_n=\frac{n-(x_{n-1}+\frac{y_{n-1}}{n})x_{n-1}}{y_{n-1}}=\frac{n-x_{n-1}^2-\frac{x_{n-1}y_{n-1}}{\sqrt{n}}}{y_{n-1}}$$

$$=\frac{x_{n-1}^2+y_{n-1}^2-x_{n-1}^2-\frac{x_{n-1}y_{n-1}}{\sqrt{n}}}{y_{n-1}}=y_{n-1}-\frac{x_{n-1}}{\sqrt{n}}$$

This completes the proof.

By iterating the endpoints of unit legs (alternatively, outer vertices of the spiral) within a spreadsheet using formulas (5), one can construct a spiral pictured in Figure 6.1. Such use of a spreadsheet illustrates "the way in which software can embody a mathematical definition" (Conference Board of the Mathematical Sciences, 2001, p. 132), in this case the recursive definition of the outer vertices of the spiral of Theodorus.

Note that the behavior of the coordinate x_n within the spiral is as follows. Starting from x_0, the coordinate first increases in absolute value until the y-coordinate becomes positive (the vertex crosses the x-axis); at that time, x_n decreases in absolute value until it passes through the origin. Then both x_n and y_n decrease until the spiral ends.

4. Parameterization of Recurrence Relation (2)

A mathematically rich and computer-enabled exploration can result from the parameterization of equation (2) in the form

$$F_n = \sqrt{aF_{n-1}^{\ 2} + b}, \quad F_1 = 1 \tag{11}$$

Here are the first five terms of sequence (11):

$$F_1 = 1, F_2 = \sqrt{a+b}, F_3 = \sqrt{a^2 + ab + b},$$
$$F_4 = \sqrt{a^3 + a^2 b + ab + b}, F_5 = \sqrt{a^4 + a^3 b + a^2 b + ab + b}.$$

In order to decide the convergence of the sequence F_n depending on parameters a and b, one has to derive a closed formula for F_n.

Proposition 2. The closed formula for the sequence F_n has the form

$$F_n = \sqrt{a^{n-1} + b\frac{1-a^{n-1}}{1-a}} \tag{12}$$

Proof. To begin note that formula (12) is undefined when $a = 1$. However, by defining $F_n \big|_{a=b=1} = \lim_{a\to 1} F_n$ and taking into account that

$$\lim_{a\to 1} \frac{1-a^{n-1}}{1-a} = \lim_{a\to 1} \frac{(1-a)(a^{n-2} + a^{n-3} + \ldots + 1)}{1-a}$$
$$= \lim_{a\to 1}(a^{n-2} + a^{n-3} + \ldots + 1) = n-1,$$

one can see that when $a = b = 1$, formula (12) is in agreement with formula (2). Although the verification of a formula for a particular case cannot be accepted as a proof, such a preliminary demonstration can serve as a motivation for a formal proof.

Formula (12) can be proved by the method of mathematical induction. Indeed, let $n = 1$. Then $F_1 = \sqrt{a^o + b\frac{1-a^0}{1-a}} = 1$. Assuming that formula (12) is true, one can make transition from n to $n+1$ as follows

$$F_{n+1} = \sqrt{a(a^{n-1} + b\frac{1-a^{n-1}}{1-a}) + b} = \sqrt{a^n + b\frac{a - a^n + 1 - a}{1-a}}$$
$$= \sqrt{a^n + b\frac{1-a^n}{1-a}}.$$

This completes the proof.

Proposition 3. When $|a| < 1$ the sequence F_n converges; otherwise the sequence F_n diverges.

Proof. Note that if $|a| < 1$, then $\lim_{n\to\infty} a^n = 0$. Assuming that there exists $\lim_{n\to\infty} F_n = l > 0$, one can pass to the limit in equation (12) to have

$$\lim_{n\to\infty} F_n = \lim_{n\to\infty} \sqrt{a^{n-1} + b\frac{1-a^{n-1}}{1-a}} = \sqrt{\frac{b}{1-a}}$$

Therefore, if $|a| < 1$, regardless of the value of b,

$$\lim_{n\to\infty} F_n = \sqrt{\frac{b}{1-a}}.$$

The last relation is in agreement with relation (1) as it follows from the latter that $\lim_{n\to\infty} f_n = \lim_{n\to\infty} \sqrt{n} = \infty$. Finally, when $|a| \geq 1$ we have $\lim_{n\to\infty} F_n = \infty$. This completes the proof.

The formal proof of convergence and/or divergence of sequence (12) can be complemented by an informal demonstration of these phenomena using the *Graphing Calculator 3.5* (*GC*) and a spreadsheet. While the use of the latter tool is rather straightforward, the use of the former tool requires the utilization of two-variable inequalities in the construction of the so-called bisector-bounded staircase that demonstrates the behavior of iterations depending on the values of parameters.

The utilization of two-variable inequalities as tools for the construction of images of points and line segments using the *GC* was already discussed in Chapters 2 and 4. Here, we need to construct the image of a segment that belongs to the vertical line $x = x^*$ located between the graphs of the functions $y = f(x)$ and $y = g(x)$, $f(x) < g(x)$. Such segment can be described as the set

$$\{(x,y) | -\varepsilon < x - x^* < \varepsilon, f(x) < y < g(x)\} \quad (13)$$

where ε is the "thickness" of the segment. The set of points defined through (13) is equivalent to the domain of the inequality

$$\sqrt{y-f(x)}\sqrt{g(x)-y}\sqrt{\varepsilon-|x-x^*|}>0 \tag{14}$$

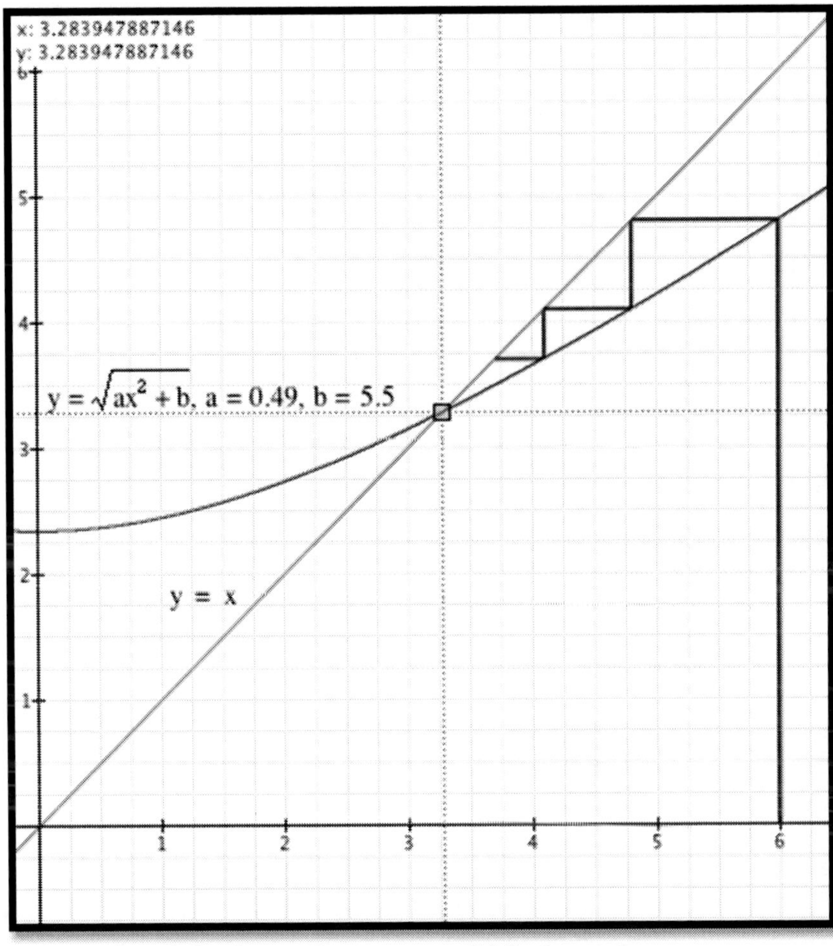

Figure 6.4. Bisector-bounded staircase of convergence of f_n for $0 < a < 1$.

Similarly, the image of a segment that belongs to the horizontal line $y = y^*$ located between the points (x_1, y^*) and (x_2, y^*), $x_1 < x_2$, can be described as the set

$$\{(x,y)\,|\,x_1 < x < x_2, -\varepsilon < y - y^* < \varepsilon\} \tag{15}$$

where ε is the "thickness" of the segment. In turn, the set of points defined through (15) is equivalent to the domain of the inequality

$$\sqrt{x-x_1}\sqrt{x_2-x}\sqrt{\varepsilon-|y-y^*|}>0 \tag{16}$$

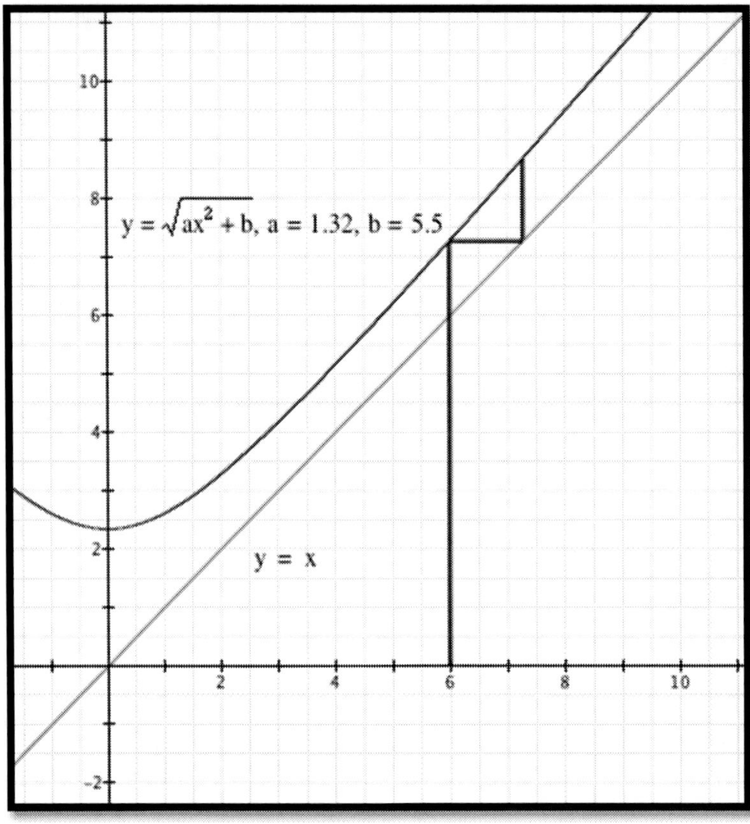

Figure 6.5. The iterations f_n diverge when $a > 1$.

Using inequalities (14) and (16) in the case of $f(x)=\sqrt{ax^2+b}$, $0<a<1$, and $g(x)=x$, the following eight inequalities

$\sqrt{y}\sqrt{\sqrt{ax^2+b}-y}\sqrt{\varepsilon-|x-6|}>0$ (the first vertical segment),

$\sqrt{6-x}\sqrt{x-\sqrt{36a+b}}\sqrt{\varepsilon-|y-\sqrt{36a+b}|}>0$ (the first horizontal segment),

$\sqrt{x-y}\sqrt{y-\sqrt{a(36a+b)+b}}\sqrt{\varepsilon-|x-\sqrt{36a+b}|}>0$ (the second vertical segment),

$\sqrt{\sqrt{36a+b}-x}\sqrt{x-\sqrt{a(36a+b)+b}}\sqrt{\varepsilon-|y-\sqrt{a(36a+b)+b}|}>0$ (the second horizontal segment),

$\sqrt{x-y}\sqrt{y-\sqrt{a(a(36a+b)+b)+b}}\sqrt{\varepsilon-|x-\sqrt{a(36a+b)+b}|}>0$ (the third vertical segment),

$$\sqrt{\sqrt{a(36a+b)+b}-x}\sqrt{x-\sqrt{a(a(36a+b)+b)+b}}$$
$$\sqrt{\varepsilon-|y-\sqrt{a(a(36a+b)+b)+b}|}>0$$

(the third horizontal segment),

$\sqrt{x-y}\sqrt{y-\sqrt{a(a(36a+b)+b)+b}}\sqrt{\varepsilon-|x-\sqrt{a(a(36a+b)+b)+b}|}>0$ (the fourth vertical segment),

$$\sqrt{\sqrt{a(a(36a+b)+b)+b}-x}\sqrt{x-\sqrt{a(a(a(36a+b)+b)+b)+b}}$$
$$\cdot\sqrt{\varepsilon-|y-\sqrt{a(a(a(36a+b)+b)+b)+b}|}>0$$

(the fourth horizontal segment), enable for the construction of the bisector-bounded staircase (Figure 6.4), which, in the case $(a, b) = (0.49, 5.5)$, starting at the point $(x,y)=(6,0)$, jumps to the point $(x,y)=(6,\sqrt{36a+b})$, and through (an infinite number of) "left-and-down" moves converges eventually to the point $(x,y)=(3.2839...,3.2839...)$. This numerical result is in agreement with Proposition 3 as $\sqrt{\frac{b}{1-a}}\Big|_{(a,b)=(0.49,5.5)}=0.3839...$. Also, the four radicands

$$36a+b,\ a(36a+b)+b,\ a(a(36a+b)+b)+b,\ a(a(a(36a+b)+b)+b)+b$$

coincide with those in the terms F_2, F_3, F_4, F_5 when $F_1 = 6$ in (11).

The same phenomenon can be observed when recurrence (11) is iterated within a spreadsheet (Figure 6.6). On the other hand, the case $a > 1$ provides a qualitatively different phenomenon when recurrent sequence (11) does not have limit and tends to infinity as shown in Figure 6.5 (a running away staircase) through the following three inequalities

$$\sqrt{y}\sqrt{\sqrt{ax^2+b}-y}\sqrt{\varepsilon-|x-6|}>0$$

(the first vertical segment),

$$\sqrt{\sqrt{36a+b}-x}\sqrt{x-6}\sqrt{\varepsilon-|y-\sqrt{36a+b}|}>0$$

(the first vertical segment),

$$\sqrt{y-x}\sqrt{\sqrt{a(36a+b)+b}-y}\sqrt{\varepsilon-|x-\sqrt{36a+b}|}>0$$

(the second vertical segment), and in Figure 6.7 through the use of a spreadsheet.

	A	B	C
1	*a*	*b*	*fn*
2	0.49	5.5	1
3			2.44744765
4			2.90432436
5			3.10373952
28			3.28394787
29			3.28394788
30			3.28394788
31			3.28394789

Figure 6.6. Convergence to $\sqrt{\frac{b}{1-a}}$, $0 < a < 1$.

	A	B	C
1	*a*	*b*	*fn*
2	1.32	5.5	1
3			2.61151297
4			3.80820168
5			4.96418855
28			157.469754
29			180.934173
30			207.890766
31			238.85982

Figure 6.7. Divergence for $a > 1$.

5. Generating Polygonal Numbers through a Sieve-Like Process

As was mentioned above, another concept that can be motivated by the spiral of Theodorus is that of polygonal numbers. These numbers can be developed through a process that resembles the Sieve of Eratosthenes[2]—an ancient method of gradual elimination of counting numbers with more than two different divisors (composite numbers) from the set of

[2] Eratosthenes of Cyrene—a Greek scholar of the 3rd century B.C.

counting numbers. Indeed, through eliminating all multiples of two, the smallest number to survive is three; then, after all multiples of three are eliminated, five is the smallest number to survive; then, after all multiples of five are eliminated, seven is the smallest number to survive; and so on.

In the case of polygonal numbers, the sieve-like process is shown in Figure 6.8 for triangular, square, pentagonal, and hexagonal numbers. To develop triangular numbers, one starts with the unity—the first triangular number; skipping one number yields three—the second triangular number, skipping two consecutive numbers yields six—the third triangular number, skipping three consecutive numbers yields ten—the fourth triangular number, and so on. In terms of operation of addition, skipping one number is equivalent to "adding two," skipping two consecutive numbers is equivalent to "adding three," and so on. In general, setting t_n to represent the n-th triangular number, the move from t_{n-1} to t_n requires skipping $n - 1$ consecutive counting numbers, or, in terms of addition, adding $(n - 1) + 1$ to t_{n-1}.

Similarly, the process of moving from the square number s_{n-1} to the square number s_n results in skipping $2(n - 1)$ numbers, or adding $2(n - 1) + 1$ to s_{n-1}; moving from the pentagonal number p_{n-1} to the pentagonal number p_n results in skipping $3(n - 1)$ numbers, or adding $3(n - 1) + 1$ to p_{n-1}; moving from the hexagonal number h_{n-1} to the hexagonal number h_n results in skipping $4(n - 1)$ numbers, or adding $4(n - 1) + 1$ to h_{n-1}. One can note that each of the coefficients in $(n - 1)$ represents the number of sides in the corresponding polygon diminished by two. The above physical descriptions can also be given geometric interpretations as one develops polygonal numbers using dot diagrams. For example, as shown in Figure 6.9, a hexagonal number is a sum of four triangular numbers, one of the same rank and three of the previous rank. More specifically, $66 = 21 + 15 + 15 + 15$ meaning that the hexagonal number of rank six is the sum of the triangular number of rank six and three triangular numbers of rank five. Alternatively, this relationship can be observed through dividing a hexagon into four triangles by connecting one of its vertices with three non-adjacent vertices.

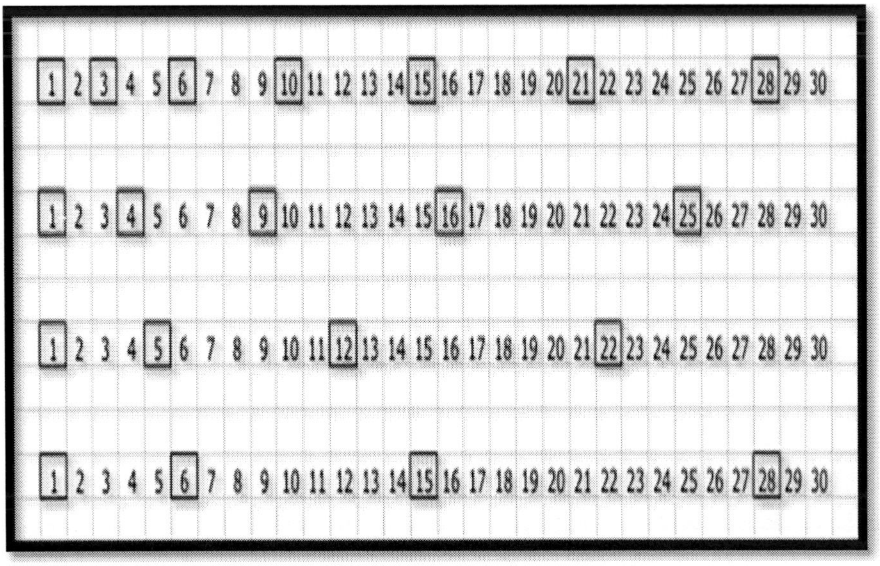

Figure 6.8. Developing polygonal numbers as a sieve process.

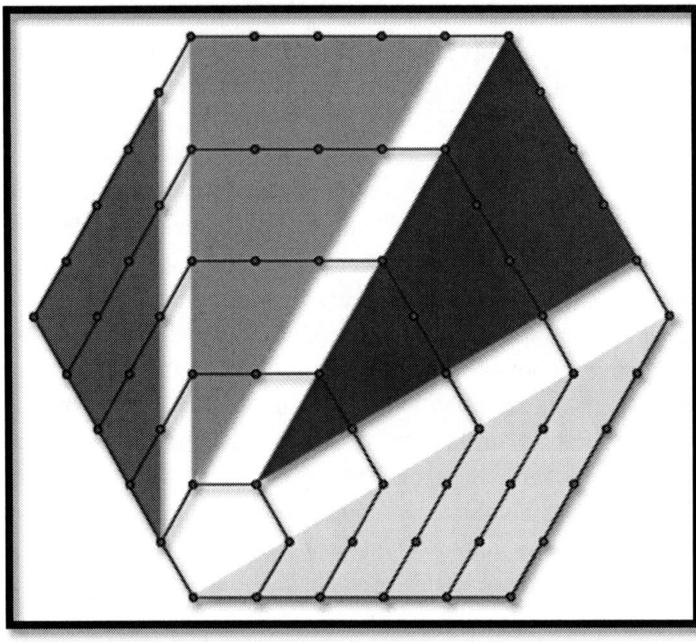

Figure 6.9. A dot diagram for the first six hexagonal numbers and its connection to triangular numbers.

Another observation that can result from Figure 6.9 is that the hexagonal number of rank six (that is, the total number of dots on the sides of the five hexagons) is the sum of the hexagonal number of rank five (the above number of dots excluding those belonging to the four largest sides not connected to the original vertex) plus five repeated four times and increased by one (the number of dots not included into the previous sum). This observation can be generalized to the dot diagram for the *m*-gonal numbers.

6. Developing Closed and Recursive Formulas for Polygonal Numbers

By setting $P(m, n)$ to represent the *m*-gonal number of rank *n* the following recursive definition can be introduced

$$P(m,n) = P(m,n-1) + (n-1)(m-2) + 1,\ P(m,1) = 1,\ m \geq 3 \qquad (17)$$

In particular, recursive definitions of triangular, square, pentagonal, and hexagonal numbers are, respectively,

$$t_n = t_{n-1} + n, t_1 = 1;$$

$$s_n = s_{n-1} + 2n - 1, s_1 = 1;$$

$$p_n = p_{n-1} + 3n - 2, p_1 = 1;$$

$h_n = h_{n-1} + 4n - 3, h_1 = 1$.

In order to derive a closed formula for $P(m, n)$, the first n cases of definition (17) can be written as follows

$$P(m,1) = 1$$

$$P(m,2) = P(m,1) + (m-2) + 1$$

$$P(m,3) = P(m,2) + 2(m-2) + 1$$

$$P(m,4) = P(m,3) + 3(m-2) + 1$$

$$P(m,n) = P(m,n-1) + (n-1)(m-2) + 1$$

Adding the above n relations yields

$$\begin{aligned} &P(m,1) + P(m,2) + P(m,3) + P(m,4) + \ldots + P(m,n) \\ &= P(m,1) + P(m,2) + P(m,3) + \ldots \\ &+P(m,n-1) + n + (m-2)(1+2+3+\ldots+(n-1)) \end{aligned}$$

Cancelling out identical terms in both sides of the last equality and adding the first $n - 1$ counting numbers by substituting $n - 1$ for n in formula (2) of Chapter 1, results in

$$P(m,n) = n + (m-2)(1+2+3+\ldots+(n-1)) = n + \frac{(m-2)(n-1)n}{2}$$

whence

$$P(m,n) = \frac{(n-1)n}{2}(m-2) + n \tag{18}$$

In particular, the sequences $t_n = \frac{n(n+1)}{2}$, $s_n = n^2$, $p_n = \frac{n(3n-1)}{2}$, $h_n = n(2n-1)$ represent, respectively, triangular ($m = 3$), square ($m = 4$), pentagonal ($m = 5$), and hexagonal ($m = 6$) numbers.

Formula (18) can be given the following interpretation: A polygonal number of rank n is a linear function of its side m with the slope $\frac{(n-1)n}{2}$ — the triangular number of rank $n - 1$— that passes through the point $(2, n)$. That is, $P(2, n) = n$, meaning that counting numbers are polygonal numbers of side two. Put another way, assuming $m \geq 2$, formula (18) defines a family of rays (depending on n) with the end point $(2, n)$ and the slope t_{n-1}. Given n, each such ray passes through the set of points with integer coordinates such that the difference between the vertical coordinates of two consecutive points is t_{n-1}. Formula (18) can be re-written in the form $P(m,n) = t_{n-1}(m-2) + n$ from where it follows that

$$P(m,n) - P(m-1,n) = t_{n-1}(m-2) + n - t_{n-1}(m-3) - n = t_{n-1},$$

that is,

$$P(m,n) = P(m-1,n) + t_{n-1} \qquad (19)$$

Formula (19) is a recursive representation of polygonal numbers in terms of side rather than in terms of rank as in the case of formula (17). For example, the pentagonal and square numbers of rank four are, respectively, 22 and 16. Their difference,

$$P(5, 4) - P(4, 4) = 22 - 16 = 6 = t_3,$$

is the triangular number of rank three. The very form of formula (19) emphasizes the meaning of slope as the vertical increment, when the horizontal increment is equal to one.

In the context of recursive definitions for polygonal numbers, one can mention the program of Axiomatization of arithmetic, originally undertaken by Grassmann—a German mathematician, physicist, and linguist of the 19th century whose work as a schoolteacher[3] delayed the recognition of his ideas (Kline, 1980). Proceeding from the recursive definition of counting numbers, Grassmann introduced the operations of addition and multiplication through recursive definitions also. This (historically significant) approach allowed all principles of arithmetic to become logical implications of the basic definitions. It is interesting to note that Grassmann has also been credited with the translation and the development of the full lexicon of the Rigveda—an ancient Indian collection of hymns dedicated to the gods in Hinduism (Klein, 1979).

7. Interpreting Summation Formulas for Polygonal Numbers in the Language of Sulvasutras

Following the ancient tradition of integrating geometry and number theory, polygonal numbers can be associated with areas produced by cords stretched in the non-unit legs of the right triangles in the spiral of Theodorus. For example, the leg of length $\sqrt{P(m,i)}$ generates the square of area $P(m,i)$—the m-gonal number of rank i. At that point, different summation problems can be posed in the context of the spiral of Theodorus. In general, problem posing is an important pedagogical tool. It was characterized by the National Council of Teachers of Mathematics (1989) as "an activity that at the heart of doing mathematics" (p. 138). In the context of problem posing teachers can develop skills in problem analysis leading to the

[3] As mentioned in Chapter 5, among schoolteachers of mathematics in the 19th century Germany were also Kummer, Weierstrass, and Worpitzky.

understanding of what makes one problem less difficult to solve than the other and what kind of problem-solving alternatives one might utilize in dealing with more difficult problems.

For example, one may pose the problem of finding $\sum_{i=1}^{n} t_i$ —the sum of the first n triangular numbers. To solve this problem, one has to proceed as follows

$$\sum_{i=1}^{n} t_i = \sum_{i=1}^{n} \frac{i(i+1)}{2} = \frac{1}{2}(\sum_{i=1}^{n} i + \sum_{i=1}^{n} i^2) = \frac{1}{2}(\frac{n(n+1)}{2} + \frac{n(n+1)(2n+1)}{6})$$
$$= \frac{n(n+1)(n+2)}{6}$$

that is

$$\sum_{i=1}^{n} t_i = \frac{n(n+1)(n+2)}{6} \qquad (20)$$

Note that in deriving the summation formula for triangular numbers we used formulas (2) and (4) developed earlier in Chapter 1 for the sums of the first n counting numbers and their squares, respectively. It turned out that although the arithmetical and geometrical structures of triangular numbers are simpler than that of square numbers, the summation of triangular numbers is based on the summation of square numbers, the latter technique already known to Babylonians in the 2nd millennium B.C. with square being the main geometric figure in Sulvasutras. On the other hand, as was discussed in Chapter 1, the exact value of the infinite sum of the reciprocals of triangular numbers is much easier to find than that of square numbers.

Formula (20) can be given the following interpretation: The sum of the first n triangular numbers is equal to one-sixth of the product of three consecutive counting numbers starting from n. This interpretation, in turn, implies that the product of three consecutive integers is a multiple of six. Alternatively, using a Sulvasutras-like language, one can say that, numerically, the sum of the volumes of n prisms six linear units tall, the bases of which are produced by cords of the lengths $\sqrt{t_i}$, i = 1, 2, 3, …, n, is equal to the volume of the prism with the dimensions n, $n + 1$, $n + 2$. In the words of Kline (1985), "The intimate connection between mathematics and objects and events in the physical world is reassuring, for it means that we can not only hope to understand the mathematics proper, but also expect physically meaningful and valuable conclusions" (pp. 32-33). However, it should be noted that had we posed a problem of partitioning such a prism into n prisms, this kind of problem would have been difficult to solve. By the same token, if another summation formula for n summands (areas or volumes) with the right-hand side represented by the product $n(n+1)(n+2)$ were found, a new mathematical connection could be established.

One can formulate the problem of finding the sum $\sum_{i=1}^{n} \frac{t_i}{i}$ (e.g., the union of certain parts of different squares in which the larger the square the smaller the corresponding part). Just by appearance, this sum is more complicated when compared to $\sum_{i=1}^{n} t_i$. Ironically, a geometrically more complicated sum is analytically easier to find. Indeed,

$$\sum_{i=1}^{n}\frac{t_i}{i}=\sum_{i=1}^{n}\frac{(i+1)}{2}=\frac{1}{2}\sum_{i=1}^{n}(i+1)$$
$$=\frac{1}{2}\sum_{j=2}^{n+1}j=\frac{1}{2}(\frac{(n+1)(n+2)}{2}-1)=\frac{n(n+3)}{4}.$$

In geometric terms, this analytic observation has the following, apparently an unanticipated, geometric interpretation: it is easier to assemble certain parts of different squares into a single square than to make a square out of a number of smaller squares.

In much the same way, considering triangles with integer hypotenuses within the spiral of Theodorus, one can arrive at the formula

$$1+\frac{4}{2}+\frac{9}{3}+...+\frac{n^2}{n}=\frac{n(n+1)}{2}$$

which can be given the following interpretation in the language of Sulvasutras: Using certain parts of squares produced by cords stretched in the legs of triangles expressed by square numbers, one can construct a square stretched by a cord in the leg of a triangle expressed by a triangular number. Interpretations of that kind reveal hidden connections between different geometric shapes. These connections are due to complex relationships that exist among real numbers.

One can formulate the general problem of finding the sum of the first n polygonal numbers multiplied by the reciprocal of their positional rank, $\sum_{i=1}^{n}\frac{P(m,i)}{i}$. To this end, multiplying the sum $1+\frac{P(m,n)}{n}$ by $\frac{n}{2}$ yields

$$\sum_{i=1}^{n}\frac{P(m,i)}{i}=\frac{n(n-1)}{4}(m-2)+n \tag{21}$$

A useful practice in making mathematical connections by moving from general to specific is to consider special cases of formula (21) for m = 6, m = 8 and m = 10, and give geometric interpretations of the formulas using the language of Sulvasutras. Furthermore, one can be encouraged to compare so obtained special cases with the formulas for s_n, p_n, and h_n. This comparison can lead to the following unexpected generalization.

Proposition 4. Every polygonal number $P(m, n)$ can be represented as a linear combination of n consecutive polygonal numbers of side $2m - 2$ as follows

$$P(m,n)=\sum_{i=1}^{n}\frac{P(2m-2,i)}{i}.$$

Proof. Replacing m by $2m-2$ in formula (21) yields

$$\sum_{i=1}^{n}\frac{P(2m-2,i)}{i}=\frac{n[(n-1)(2m-4)+4]}{4}$$
$$=\frac{n(n-1)(m-2)+2n}{2}=\frac{n(n-1)}{2}(m-2)+n.$$

Reference to formula (18) completes the proof.

Remark 1. Proposition 4 means that any triangular number can be represented as a linear combination of consecutive square numbers; any square number can be represented as a linear combination of consecutive hexagonal numbers, any pentagonal number can be represented as a linear combination of consecutive octagonal numbers, and so on.

Finally, one can pose a problem of finding the sum of the first n polygonal numbers of side m.

Proposition 5. If $P(m, i)$ is the polygonal number of side m and rank i, then

$$\sum_{i=1}^{n}P(m,i)=\frac{n(n^2-1)}{6}(m-2)+\frac{n(n+1)}{2} \qquad (22)$$

Proof. Using formulas (18) and (20) one can write

$$\sum_{i=1}^{n}P(m,i)=\sum_{i=1}^{n}\left(\frac{i(i-1)}{2}(m-2)+i\right)$$
$$=(m-2)\sum_{i=1}^{n-1}t_i+\sum_{i=1}^{n}i=(m-2)\frac{n(n-1)(n+1)}{6}$$
$$+\frac{n(n+1)}{2}=\frac{n(n^2-1)}{2}(m-2)+\frac{n(n+1)}{2}.$$

Remark 2. In the right-hand side of formula (22), the coefficient $\frac{n(n-1)(n+1)}{6}$ in $(m-2)$ represents the so-called tetrahedral number of rank $n-1$ for which the notation T_{n-1} can be used.

Remark 3. When $m = 3$, formula (22) yields

$$\sum_{i=1}^{n}t_i=\frac{n(n^2-1)}{6}+\frac{n(n+1)}{2}=T_{n-1}+t_n$$

On the other hand, due to formula (20)

$$\sum_{i=1}^{n} t_i = \frac{n(n+1)(n+2)}{6} = \frac{(n+1)[(n+1)-1][(n+1)+1]}{6}$$

$$= \frac{(n+1)[(n+1)^2-1]}{6} = T_n$$

That is,

$$T_n = T_{n-1} + t_n,\ T_1 = 1 \qquad (23)$$

Difference equation (23) may be construed as a recursive definition of tetrahedral numbers. Geometrically, tetrahedral numbers, being a three-dimensional analogue of triangular numbers, represent discrete points placed in the configuration of a triangular pyramid (tetrahedron). Tetrahedral numbers, along with triangular numbers, can be found in Pascal's triangle (Chapter 5). It is interesting to note that Pascal's triangle is referred to in China as Yanghui triangle as it had been known to the 13th century Chinese mathematician Yanghui (Weisstein, 1999).

8. The Spiral Motivates Transition from Summation to Estimation

Whereas for certain sums closed formulas are difficult to find (compare $\sum i$ to $\sum t_i$), there are sums for which such formulas either do not exist or their finding may require efforts not worthy of the result. In such cases, one can reduce the problem of summation to its alternative—the problem of estimation. Put another way, one can move from the world of equalities to the world of inequalities. To this end, the spiral of Theodorus can motivate the evaluation of the linear combinations of areas of right triangles (rather than squares). Because the triangles do not necessarily have rational areas, this case is different from the one explored in the previous section.

To begin note that one can associate the non-unit leg of each triangle with its positional rank within the spiral of Figure 6.1. Four triangles with an interesting property can be identified. Setting A_r to represent twice the area of triangle of positional rank r^2 (i.e., $A_r = r$), one can discover that

$$A_1 + \frac{A_2}{2} + \frac{A_3}{3} + \frac{A_4}{4} = A_4 .$$

Indeed, whereas the right-hand side of the last relation is equal to four, each of the four summands in its left-hand side is equal to one. Extending the spiral to n^2 triangles, results in the equality $\sum_{i=1}^{n} \frac{A_i}{i} = n$ because each term of the sum is equal to one.

This discovery (although having a trivial algebraic interpretation), in turn, motivates the following question regarding the extended spiral: *Are there other linear combinations of areas that add up to twice the area of the largest triangle in a combination?* Compare, for example, the sum $A_1 + \frac{A_{\sqrt{2}}}{2} + \frac{A_{\sqrt{3}}}{3} + \frac{A_{\sqrt{4}}}{4}$ to $A_{\sqrt{4}}$. A simple calculation shows that this comparison does not result in equality. Rather, it yields the inequality $1 + \frac{\sqrt{2}}{2} + \frac{\sqrt{3}}{3} + \frac{\sqrt{4}}{4} > \sqrt{4}$, which, by using numerical evidence provided by a spreadsheet, can be extended over the whole spiral to get $\sum_{i=1}^{16} \frac{\sqrt{i}}{i} > \sqrt{16}$.

One may wonder as to how accurate the estimation from below by $\sqrt{16}$ is? That is, how can one estimate the sum not only from below but from above as well? Numerical evidence provided by a spreadsheet prompts the following estimate $\sum_{i=1}^{16} \frac{\sqrt{i}}{i} < 2\sqrt{16}$. Combining the two estimates yields

$$2\sqrt{16} > \sum_{i=1}^{16} \frac{\sqrt{i}}{i} > \sqrt{16} \qquad (23)$$

and geometric interpretation of inequalities (23) in the language of Sulvasutras:

> The sum of the ratios of the lengths of the cords stretched in the16 consecutive legs of the spiral that share the center of the spiral as the common point to the legs' corresponding positional ranks is greater than the length of the longest cord involved in this sum and is smaller than twice this length.

Finally, by generalizing from inequalities (23), one can come up with

Proposition 6. For any integer $n \geq 2$, the following inequalities hold true

$$\sum_{i=1}^{n} \frac{\sqrt{i}}{i} > \sqrt{n} \qquad (24)$$

$$\sum_{i=1}^{n} \frac{\sqrt{i}}{i} < 2\sqrt{n} \qquad (25)$$

Remark 4. In comparison with Propositions 4 and 5, the meaning of Proposition 6 is that it allows one to replace summation by estimation when the former operation cannot be carried out easily. Figure 6.10 shows the results of spreadsheet modeling of inequalities (24) and (25) for $n \leq 250$. One can see that as n grows larger, the estimate from above (inequality (25)) appears to be much more accurate than the estimate from below (inequality (24)).

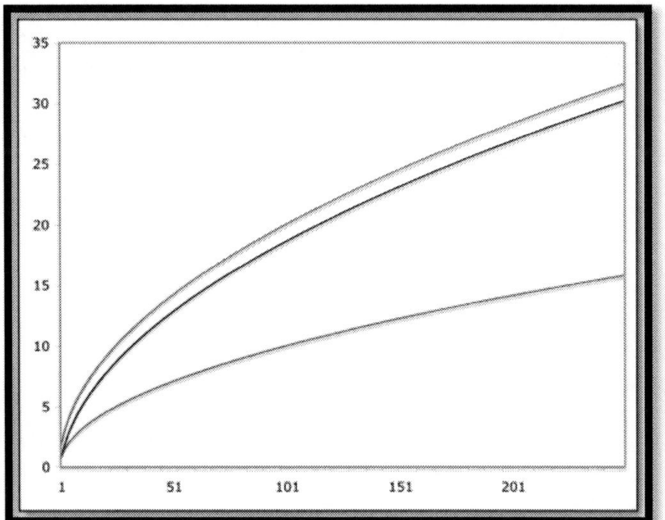

Figure 6.10. Graphical modeling of inequalities (24) and (25) using a spreadsheet.

9. Proof of Proposition 6 as an Agency for Problem Posing

Discovered by induction, Proposition 6 has to be proved formally. In mathematics teacher education, formal proof can serve multiple goals. These goals include the development of taste for rigor, fostering formal reasoning and logical thinking, connecting different (often seemingly unrelated) concepts, discovering multiple ways of arriving to the same conclusion, and exploring *why* a statement that one proves is true (Hersh, 1993). Below, both the proof and its didactic potential for an extended learning experience in problem posing will be discussed. In part for the sake of brevity, this discussion will mostly focus on the proof of inequality (24). Yet, in connection with the (computationally established) fact that inequality (25) is a more accurate estimate of the sum in question than inequality (24), it can be shown that the former requires a slightly more complicated proof technique than the latter.

Doing proof as an intellectual activity may elicit mathematical behavior and can stimulate productive thinking. The very context of proving can motivate activities that go beyond the narrow purpose of the validation of conjectures. It is in this sense that proving may be given an agency that motivates problem posing, "a platform from which further development proceeds" (Davis, 1985, p. 23). In what follows, mathematical induction proof of inequality (24) will be used as an agency for problem posing.

9.1. Revisiting Geometry through Verifying Base Clause

The first step of the demonstrative phase is to establish the base clause by showing that inequality (24) is true for $n = 2$. More specifically, one has to verify that

$$1+\frac{\sqrt{2}}{2}>\sqrt{2} \qquad (26)$$

Inequality (26), when approached from a conceptual rather than pure numerical perspective and situated in the context of geometry, enables one to revisit a number of geometric propositions associated with it. This approach is both conceptual and contextual as geometry serves as a context for proof of an abstract statement developed through interpreting geometric constructions in analytical terms. Indeed, a number of problematic situations can emerge through the multiplicity of ways used to verify inequality (26).

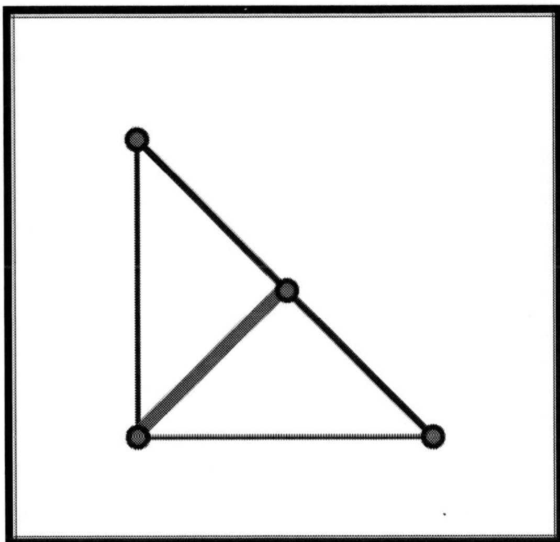

Figure 6.11. A geometric interpretation of inequality (26).

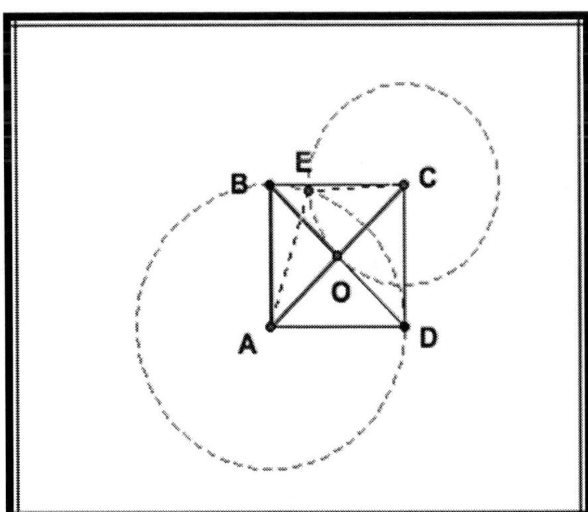

Figure 6.12. Revealing hidden meaning of inequality (26).

Step 1. Prove inequality (26) by noting that a cord stretched in a leg of an isosceles right triangle is bigger than the one stretched in the altitude dropped on the hypotenuse (Figure 6.11).

Proof. It follows from Figure 6.11 that $1 > \frac{\sqrt{2}}{2}$ whence $1 + \frac{\sqrt{2}}{2} > \frac{\sqrt{2}}{2} + \frac{\sqrt{2}}{2} = \sqrt{2}$.

Variation of step 1. Prove inequality (26) by interpreting it as a possibility of constructing a triangle using cords stretched along a side, diagonal and radius of square.

Proof. As shown in Figure 6.12, the interpretation of inequality (26) as the triangle inequality is *hidden* within the square ABCD and can be revealed through the auxiliary construction of the triangle AEC.

9.2. Posing Problems in the Context of Inductive Transfer

The next step is to test the transition from n to $n + 1$. In other words, assuming that inequality (24) holds true for n (inductive assumption), one has to show that after replacing n by $n + 1$ it remains true; that is,

$$\sum_{i=1}^{n+1} \frac{\sqrt{i}}{i} > \sqrt{n+1} \tag{27}$$

This transition can be demonstrated in multiple ways allowing for new mathematical activities to be introduced.

Step 2. Prove a transition from n to $n + 1$ by using a relationship between a number and its square root.

Proof. As was mentioned above, any positive number smaller than one does not exceed its square root. Therefore, a true inequality $\frac{n}{n+1} < 1$ implies that

$$\sqrt{\frac{n}{n+1}} > \frac{n}{n+1} \tag{28}$$

for $n = 1, 2, 3, \ldots$. Inequality (28) enables one to replace the difference $\sqrt{\frac{n}{n+1}} - \frac{n}{n+1}$ by zero in the following estimations from below:

$$\sum_{i=1}^{n+1} \frac{\sqrt{i}}{i} = \sum_{i=1}^{n} \frac{\sqrt{i}}{i} + \frac{\sqrt{n+1}}{n+1} > \sqrt{n} + \frac{\sqrt{n+1}}{n+1}$$
$$= \sqrt{n+1}\left(\sqrt{\frac{n}{n+1}} - \frac{n}{n+1} + 1\right) > \sqrt{n+1}.$$

Variation of step 2. Demonstrate the transition from n to $n + 1$ in inequality (24) by proving the inequality

$$\sqrt{n}+\frac{1}{\sqrt{n+1}}>\sqrt{n+1} \tag{29}$$

Proof. Multiplying both sides of inequality (29) by $\sqrt{n+1}$ results in the inequality $\sqrt{n^2+n}>n$ which is true for all $n > 0$. This completes the proof of Proposition 6.

Remark 5. One can transform inequality (29) to the form

$$\sqrt{n+1}-\sqrt{n}<\frac{1}{\sqrt{n+1}} \tag{30}$$

Inequality (30) indicates that its left-hand side gets smaller and smaller as n increases for it remains smaller than $\frac{1}{\sqrt{n+1}}$—a fraction that tends to zero as n increases. In particular, this proves that the function $f(n)=\sqrt{n+1}-\sqrt{n}$ tends to zero as $n\to\infty$ (see Pólya, 1965, p. 49). A related exploration can be provided by

Proposition 7. The inequality

$$\sqrt{n+1}-\sqrt{n}>\frac{1}{n+1} \tag{31}$$

holds true for all $n \geq 3$. In other words, no matter how small the difference between $\sqrt{n+1}$ and $\sqrt{n}$ is, it does exceed the value of $\frac{1}{n+1}$ beginning from $n = 3$.

Proof. Unlike the case of inequality (29), the proof of inequality (31) suggested below is rather intricate. It is based on the joint use of algebra of polynomials, differential calculus, and graphing technology. Indeed, by rewriting inequality (31) in an equivalent form

$$\sqrt{n+1}-\frac{1}{n+1}>\sqrt{n} \tag{32}$$

and squaring both sides of (32), yields the inequality

$$1-\frac{2}{\sqrt{n+1}}+\frac{1}{(n+1)^2}>0 \tag{33}$$

Substituting $\frac{1}{\sqrt{n+1}}$ for x in inequality (33) results in

$$x^4 - 2x + 1 > 0 \tag{34}$$

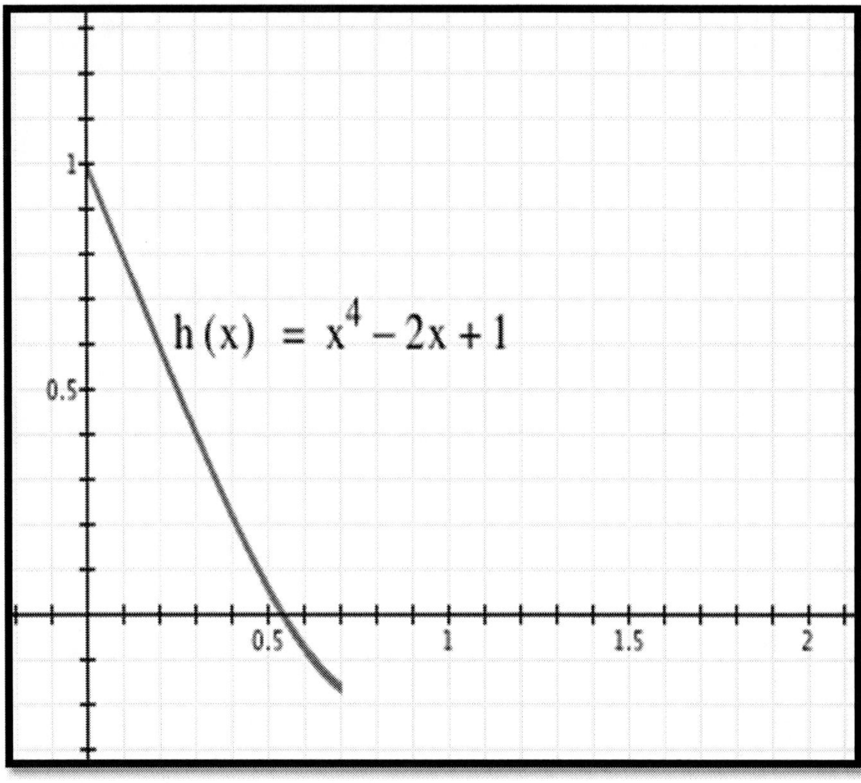

Figure 6.13. The graph of $h(x)$ in the interval $0 \leq x \leq \frac{1}{\sqrt{2}}$.

Consider the function $h(x) = x^4 - 2x + 1$ for $0 \leq x \leq \frac{1}{\sqrt{2}}$, which graph, shown in Figure 6.13, can be constructed by using the *GC*. Similar to the use of the tool discussed in section 4 of this chapter, graphing the inequality

$$\sqrt{\varepsilon - |y-(x^4-2x+1)|}\sqrt{x}\sqrt{\frac{1}{\sqrt{2}}-x} > 0$$

for a relatively small value of ε (in Figure 6.13, $\varepsilon = 0.01$) yields a curve that represents the graph of $h(x)$ in the interval $0 \leq x \leq \frac{1}{\sqrt{2}}$. As $h(0) > 0$ and $h(\frac{1}{\sqrt{2}}) < 0$, there exists x_0, $0 < x_0 < \frac{1}{\sqrt{2}}$, such that $h(x_0) = 0$. Furthermore, the derivative $h'(x) = 4x^3 - 2 < 0$ for $x < x_0 < \frac{1}{\sqrt{2}} < \frac{1}{\sqrt[3]{2}}$ implying that $h(x) > h(x_0) = 0$ for $x < x_0$. When $n \geq 3$, one has $x \leq 1/2$ and $h(1/2) > 0$. The fact that inequality (34) is true for $x \leq 1/2$ implies that inequality (31) is true for $n \geq 3$. This completes the proof.

10. The Harmonic Series Revisited

The didactic importance of inequality (31) is that it can be used in demonstrating the divergence of the series $\sum_{n=1}^{\infty}(\sqrt{n+1}-\sqrt{n})$ along with the harmonic series $\sum_{n=1}^{\infty}\frac{1}{n}$, whereas $\lim_{n\to\infty}(\sqrt{n+1}-\sqrt{n})=0$.

Proposition 8. Prove that the series $\sum_{n=1}^{\infty}(\sqrt{n+1}-\sqrt{n})$ diverges.

Proof. Using inequality (31) allows one to proceed as follows

$$\sum_{n=1}^{\infty}(\sqrt{n+1}-\sqrt{n})>\sum_{n=3}^{\infty}(\sqrt{n+1}-\sqrt{n})>\sum_{n=3}^{\infty}\frac{1}{n+1}=\sum_{n=4}^{\infty}\frac{1}{n}=\infty$$

In connection with the harmonic series, a number of historically significant references can be made also. As mentioned in section 6.3 of Chapter 1, D'Oresme was the first to demonstrate the divergence of the harmonic series. Mengoli[4], Jacob Bernoulli (already mentioned in Chapters 1 and 3), and Leibniz[5] have also been credited with the demonstration of the convergence of the harmonic series. In the 18th century, Euler discovered that the function $g(n)=1+\frac{1}{2}+\frac{1}{3}+...+\frac{1}{n}-\log(n)$ has a finite limit C = 0.5772157 … (commonly referred to as Euler's constant) thereby demonstrating that partial sums of the harmonic series behave as $\log(n)$. By using a spreadsheet, one can reach g(6,289,700) = 0.57721574439662 starting from $g(1) = 1$ in approximately ten minutes of computing on Macintosh OS X. More details about spreadsheet-based experimental approaches to investigating the behavior of sequences and series, including a visual demonstration of the divergence of the harmonic series and calculation of Euler's constant, can be found elsewhere (Abramovich, 1995).

Remark 6. The mathematical induction proof of inequality (24) discovered inductively in the context of measurement can be used as a learning environment, which extends beyond the immediate purpose of proof. Within such an environment one can come to appreciate that whereas "algebraic notation provides a kind of universal language for representing quantitative information, geometric shapes provide visual representations of ideas" (Conference Board of the Mathematical Sciences, 2001, p. 132). By using proof as a means for mathematical learning, one can explore the interplay between algebra, functions, and geometry.

---

4 Pietro Mengoli—an Italian mathematician of the 17th century.

5 Gottfried Wilhelm Leibniz (1646-1716)—a German mathematician, physicist, logician, and philosopher, one of the developers of differential and integral calculus.

11. Activity Set

1. Consider a problem about the area of a rectangle with whole number sides expressed by the Babylonians in the following rhetorical form (Van der Waerden, 1961, p. 63): "*I have multiplied length and width, thus obtaining the area. Then I added to the area the excess of the length over the width ... 183 was the result ... Required length, width and area.*"

1.a. Construct a spreadsheet to numerically model solutions to this problem.

1.b. Show that the smaller the excess of the length over the width, the bigger is the area of the rectangle.

1.c. How many different rectangles are there?

1.d. Which rectangle has the largest area? Which rectangle has the smallest area?

1.e. Which rectangle has the largest perimeter? Which rectangle has the smallest perimeter?

1.f. Is rectangle with the smallest perimeter unique?

1.g. Why there may not be more than two integer-sided rectangles with the smallest perimeter?

1.h. Make the number 183 a parameter n. What is special about rectangle whose area is equal to n square units? Find at least one such value of n.

1.i. Find rectangle with the smallest perimeter in the case $n = 281$. Is such rectangle unique? If so, what is special about this rectangle?

1.j. Extend the problem situation to include rectangles with non-integer side lengths and n being a parameter. Use the Arithmetic Mean—Geometric Mean inequality (Chapter 3) to find the rectangle with the smallest perimeter. What is the area of the rectangle with the smallest perimeter? What is the excess of the length over the width for the rectangle with the smallest perimeter? Is this excess equal to n like in the original Babylonian problem?

Chapter 7

COMPUTATIONAL EXPERIMENTS AND FORMAL DEMONSTRATION IN TRIGONOMETRY

Surprisingly, trigonometric functions proved to be admirably suited for the study of sound, electricity, radio, and a host of other oscillatory phenomena.
— Kline (1985, p.417)

1. INTRODUCTION

The Conference Board of the Mathematical Sciences (2001), in the context of recommendations for teacher preparation in geometry, emphasized the need for courses that help teachers develop "understanding of trigonometry from a geometric perspective and skill in using trigonometry to solve problems" (p. 41). The Board went on to suggest that a capstone course "is a natural place to re-examine trigonometric ideas ... and to make or reinforce connections with geometry" (ibid, p. 132). This chapter will show how these recommendations can be addressed by connecting trigonometry to geometry through the comparison of different forms of answers to a single trigonometric equation. In addition, this chapter will demonstrate how technology, including thc *Graphing Calculator 3.5* (*GC*) and *Maple* can be used to support computational experiments in trigonometry through which a variety of trigonometric equations with multiple series of roots can be formulated and solved.

The need for comparing answers in trigonometric equations emerges when the diversity of students' mathematical thinking, encouraged by current standards for teaching mathematics (National Council of Teachers of Mathematics, 2000), leads to their use of different solution strategies. In turn, different strategies might yield different, yet equivalent forms of answers in terms of arc functions. This situation is rather peculiar and in a sense is trigonometry-specific, although the notion of equivalence is known as one of the central themes of mathematics already studied at the upper elementary level (ibid, p. 144).

Typically, when we say that a problem has more than one correct answer, we mean that there exist multiple ways of solving the problem or arriving at the same answer. The multiplicity of solutions in such cases serves as a confirmation of the correctness of answer. These may include various proof techniques or different solution strategies. To clarify, consider the problem of finding two numbers with the sum 10 and the largest product. Setting $x + y = 10$, the product xy can be written as $f(x) = x(10 - x)$. There are at least three ways to find the largest value of $f(x)$.

1. Completion of square:
   $f(x) = -(x^2 - 10x + 25) + 25 = 25 - (x-5)^2, \max_{0<x<10} f(x) = 25$.
2. Differentiation:
   $f'(x) = 10 - 2x = 2(5 - x),$
   $f'(5) = 0, f'(5-\varepsilon) > 0, f'(5+\varepsilon) < 0, \max_{0<x<10} f(x) = 25.$
3. Using the Arithmetic Mean—Geometric Mean inequality (Chapter 3):
   $f(x) = x(10-x) \le (\frac{x+10-x}{2})^2 = 25, \max_{0<x<10} f(x) = 25$.

Note that in all three cases the answer is the same in terms of its symbolic representation: when $x = y = 5$ the product $x \cdot y = 25$ is the largest one.

Unlike the above numeric example, in trigonometric equations, the form of the answer, that is, the form of notation used to represent a series of roots (a solution), often depends on the problem-solving strategy chosen. Recall that most analytically solvable trigonometric equations are reducible to one of the three basic forms and their *equivalent* representations through arc functions

$$\sin x = p \Leftrightarrow x = (-1)^n \arcsin p + \pi n, \; |p| \le 1 \tag{1}$$

$$\cos x = p \Leftrightarrow x = \pm \arccos p + 2\pi n, \; |p| \le 1 \tag{2}$$

$$\tan x = p \Leftrightarrow x = \arctan p + \pi n \tag{3}$$

Hereafter, *n* denotes any integer and

$$\arcsin p \in [-\frac{\pi}{2}, \frac{\pi}{2}], \arccos p \in [0, \pi], \arctan p \in (-\frac{\pi}{2}, \frac{\pi}{2}).$$

Relations (1)-(3) can be derived using their graphic/geometric representations shown in Figures 7.1-7.3, from which it also follows that

$$\arcsin(-p) = -\arcsin p \tag{4}$$

$$\arccos(-p) = \pi - \arccos p \tag{5}$$

$$\arctan(-p) = -\arctan p \tag{6}$$

Finally, any point on the unit circle resulted from the rotation of the point (1, 0) around the origin by angle *x*, has the coordinates $(\cos x, \sin x)$ and therefore,

$$\sin^2 x + \cos^2 x = 1 \tag{7}$$

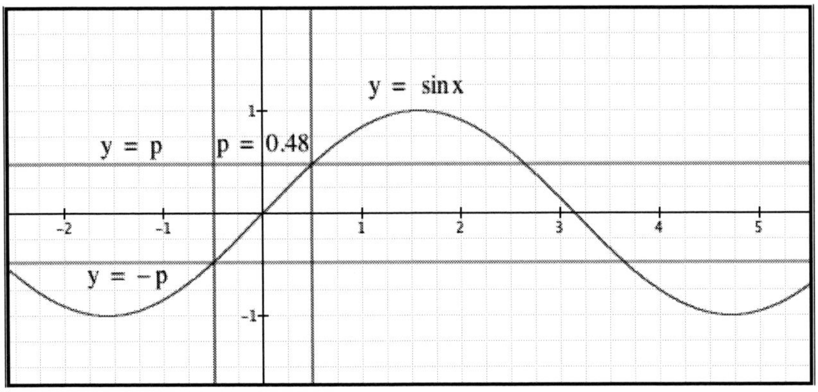

Figure 7.1. Solving the equation $\sin x = p$.

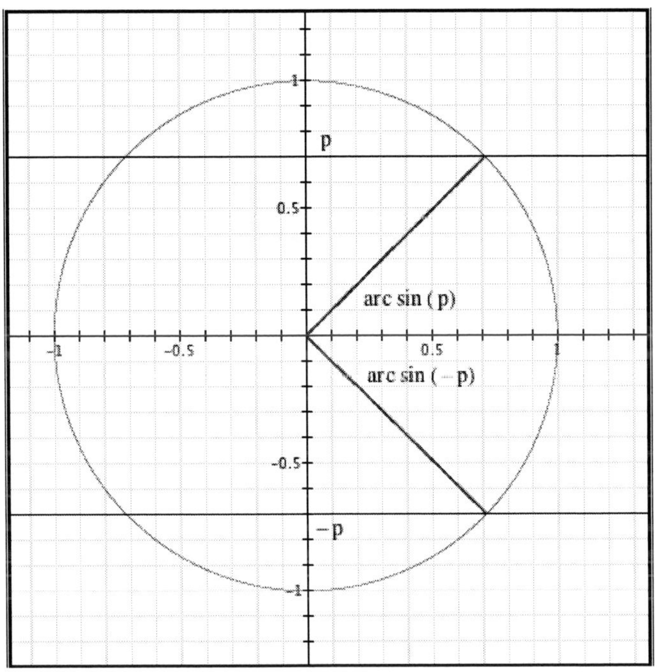

Figure 7.2. Demonstrating the identity $\arcsin(-p) = -\arcsin p$.

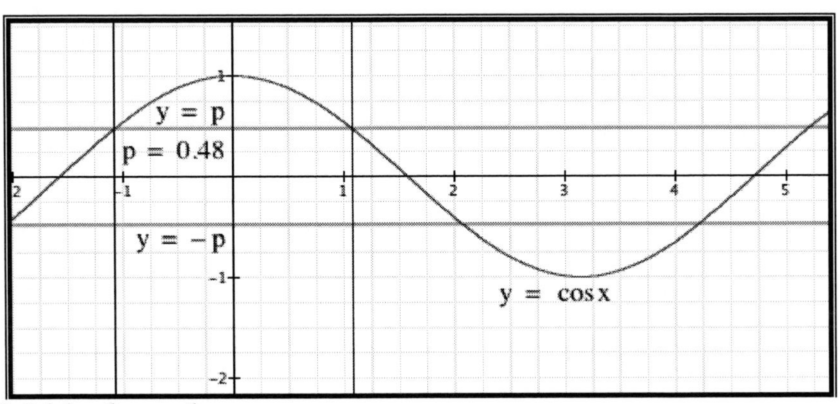

Figure 7.3. Solving the equation $\cos x = p$.

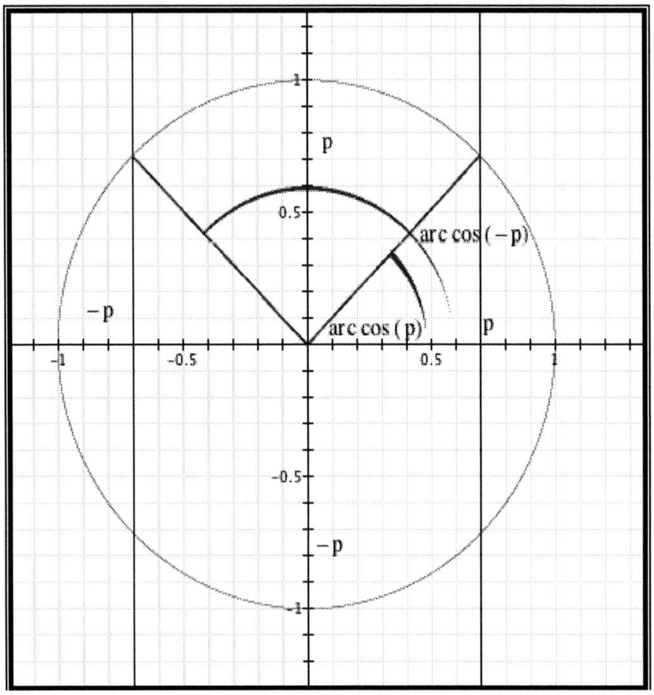

Figure 7.4. Demonstrating the identity $\arccos(-p) = \pi - \arccos p$.

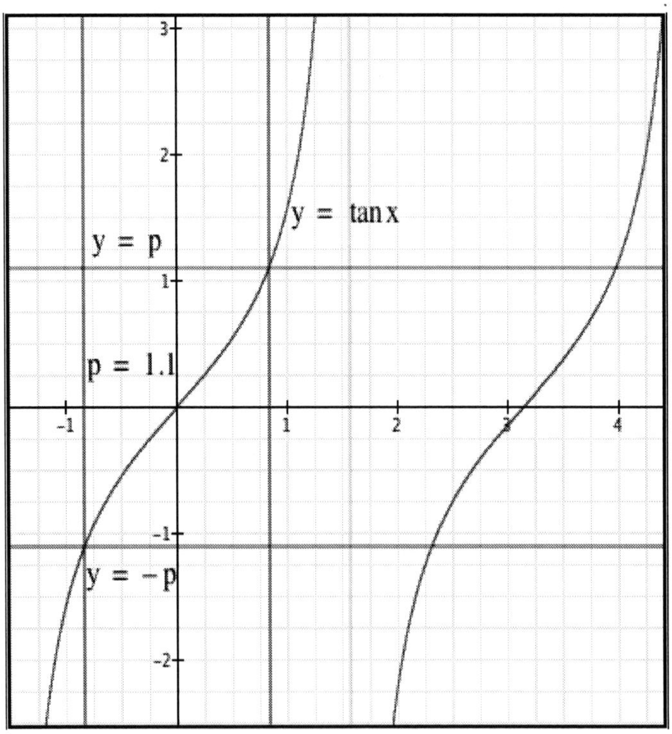

Figure 7.5. Solving the equation tan $x = p$.

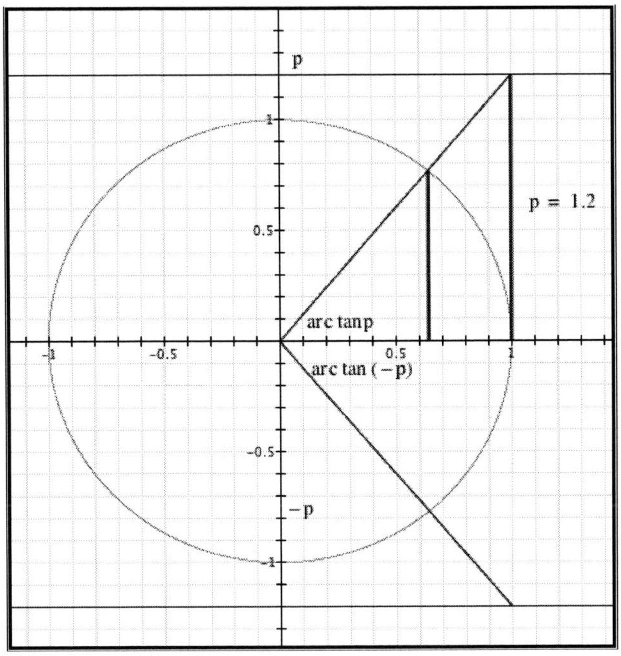

Figure 7.6. Demonstrating the identity $\arctan(-p) = -\arctan p$.

2. One Equation—Four Solutions

Consider the equation

$$2 + \cos^2 2x = (2 - \sin^2 x)^2 \tag{8}$$

which will serve as background for the ideas discussed below. Whereas equation (8) is not difficult to solve, multiple solution strategies that it affords allow the discussion to be extended into geometry and to include the appropriate use of technology.

Solution 1: Reduction to a quadratic equation in terms of $\sin^2 x$.

By using the formula

$$\cos 2x = 1 - 2\sin^2 x \tag{9}$$

equation (8) can be transformed to

$$2 + (1 - 2\sin^2 x)^2 = (2 - \sin^2 x)^2 .$$

The substitution $z = \sin^2 x$, $0 \le z \le 1$, yields

$$2 + 1 - 4z + 4z^2 = 4 - 4z + z^2$$

whence $z^2 = \frac{1}{3}$ or $z = \frac{1}{\sqrt{3}}$ (rejecting $z = -\frac{1}{\sqrt{3}}$). Therefore, $\sin^2 x = \frac{1}{\sqrt{3}}$ whence

$$\sin x = \pm\frac{1}{\sqrt[4]{3}},$$

Using formulas (1) and (4) yields

$$x = \pm\arcsin(\frac{1}{\sqrt[4]{3}}) + \pi n \qquad (10)$$

Solution 2: Reduction to a quadratic equation in terms of cos²x.
By using the formula

$$\cos 2x = 2\cos^2 x - 1, \qquad (11)$$

which follows from formulas (7) and (9), and setting $z = \cos^2 x,\ 0 \le z \le 1$, equation (8) can be transformed to the quadratic equation .

$$3z^2 - 6z + 2 = 0$$

whence

$$z_1 = \cos^2 x = \frac{3-\sqrt{3}}{2} \text{ and } z_2 = \cos^2 x = \frac{3+\sqrt{3}}{2}.$$

The latter equation is an extraneous one as $\frac{3+\sqrt{3}}{2} > 1$. Therefore,

$$\cos x = \pm\sqrt{\frac{3-\sqrt{3}}{2}}.$$

The use of formula (2) yields the series $x = \pm\arccos(\pm\sqrt{\frac{3-\sqrt{3}}{3}}) + 2\pi n$, which, as it follows from formula (5) and Figure 7.2, can be reduced to its equivalent form

$$x = \pm\arccos\sqrt{\frac{3-\sqrt{3}}{3}} + \pi n \qquad (12)$$

One can see that series (10) and (12) look quite different.
Solution 3: Reduction to a quadratic equation in terms of cos2x.

By using formula (7), equation (8) can be replaced by the equation

$$2+\cos^2 2x=(1+\cos^2 x)^2 \tag{13}$$

Then, using formula (11) in the form $\cos^2 x=\frac{1+\cos x}{2}$ and setting $z=\cos 2x$, $|z|\le 1$, equation (13) can be reduced to the quadratic equation

$$3z^2-6z-1=0$$

whence

$$z_1=\frac{3-2\sqrt{3}}{3} \text{ and } z_2=\frac{3+2\sqrt{3}}{3}.$$

Noting that $z_2>1$ leaves us with the (single) equation

$$\cos 2x=\frac{3-2\sqrt{3}}{3}$$

which, due to formula (2), is equivalent to

$$x=\pm\frac{1}{2}\arccos\frac{3-2\sqrt{3}}{3}+\pi n \tag{14}$$

Once again, the form of series (14) is quite different from (10) and (12).

Solution 4: Reduction to a quadratic equation in terms of tan²x.

Using formula (7), equation (8) can be transformed to the form

$$\begin{aligned}&2(\sin^4 x+2\sin^2 x\cos^2 x+\cos^4 x)+(\cos^2 x-\sin^2 x)^2\\&=(\sin^2 x+2\cos^2 x)^2\end{aligned} \tag{15}$$

Dividing each term of equation (15) by $\cos^4 x$ (note that $\cos x=0$ is not a solution of (15), thus the operation of division does not result in the loss of solutions) yields

$$2\tan^4 x-2\tan^2 x-1=0$$

whence (noting that $\tan^2 x>0$)

$$\tan^2 x=\frac{1+\sqrt{3}}{2}$$

or

$$\tan x = \pm\sqrt{\frac{1+\sqrt{3}}{2}}$$

Using formula (3) yields

$$x = \pm\arctan\sqrt{\frac{1+\sqrt{3}}{2}} + \pi n \qquad (16)$$

Our goal is to show the equivalence of series (10), (12), (14), and (16) in the sense that they generate identical sets of points on the number line.

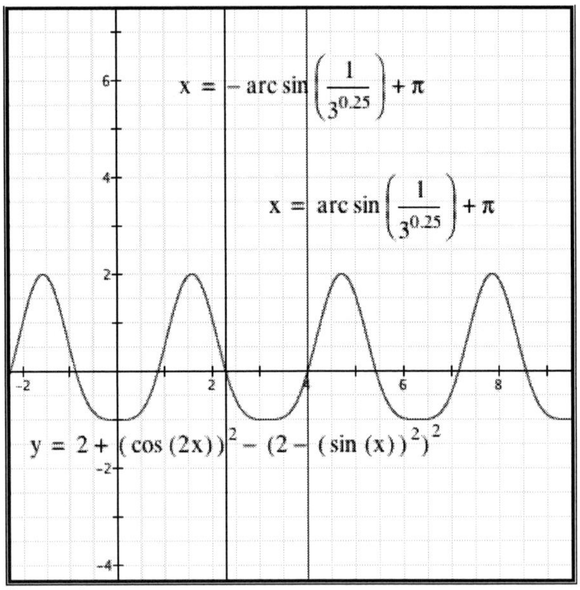

Figure 7.7. Series (10) for $n = 1$.

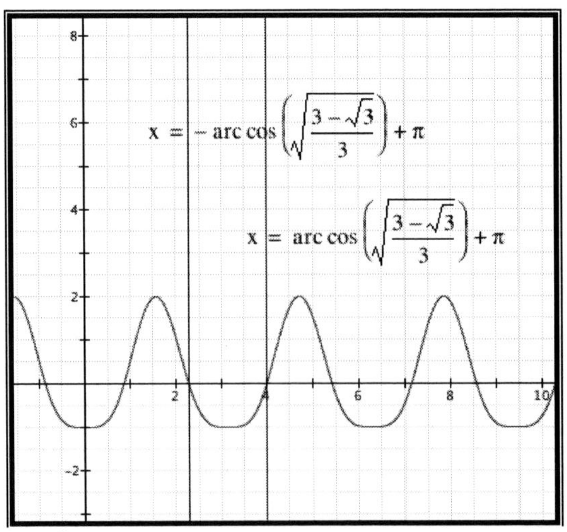

Figure 7.8. Series (12) for $n = 1$.

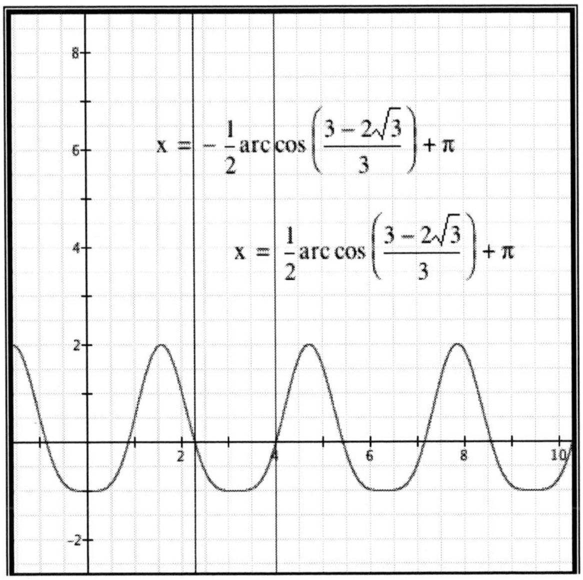

Figure 7.9. Series (14) for $n = 1$.

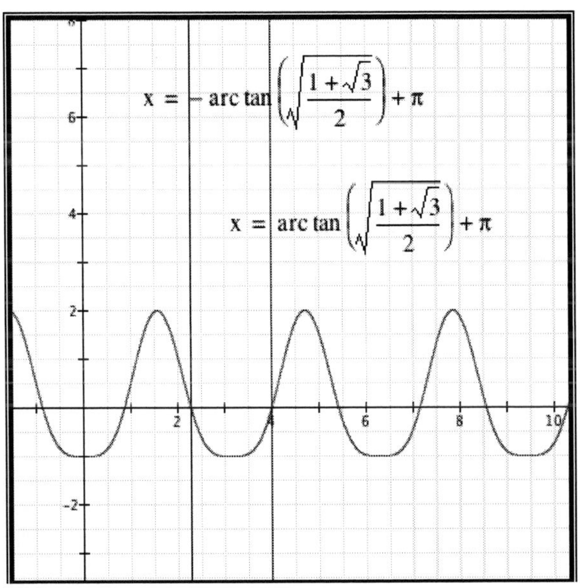

Figure 7.10. Series (16) for $n = 1$.

3. A Computer-Supported Graphical Demonstration

An appropriate form of technology that can be used effectively in checking the results through graphing is the *GC*. Its operational capability to graph functions (and, more generally, relations) depending on parameters, already employed in the previous chapters, can now be utilized in the context of graphing equation (8) concurrently with series (10), (12), (14), and (16), and setting n as a slider-controlled parameter. The graphs shown in Figures 7.7-7.10

indicate that for $n = 1$ all roots that belong to any segment on the x-axis having length π—the period of the function $y = 2 + \cos^2 2x - (2 - \sin^2 x)^2$—and generated by series (10), (12), (14), and (16) are identical regardless of a notation used to represent them. Furthermore, one can play the slider to see that this identity remains true for other integer values of n.

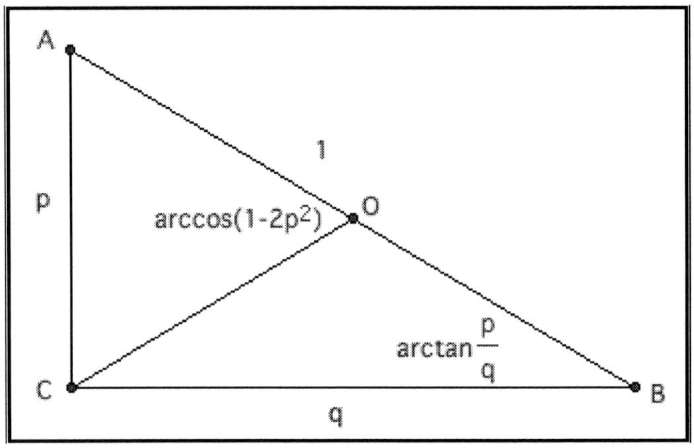

Figure 7.11. Right triangle with unit hypotenuse and arc functions.

4. A Formal Geometric Demonstration

The following two recommendations by Pólya (1973) to mathematical problem solvers are worth citing in connection with establishing the identity of different forms of solution of equation (8). The first recommendation concerns the power of geometrization: "To find a lucid geometric representation for your non-geometrical problem could be an important step toward the solution" (p. 108). The second recommendation asserts, "Generalization may be useful in the solution of problems" (p. 108). With this in mind, one can attempt to generalize the chain of equalities

$$\arcsin\frac{1}{\sqrt[4]{3}} = \arccos\sqrt{\frac{3-\sqrt{3}}{3}}$$
$$= \frac{1}{2}\arccos\frac{3-2\sqrt{3}}{3} = \arctan\sqrt{\frac{1+\sqrt{3}}{2}} \qquad (17)$$

enabling a representation associated with a right triangle ΔABC with the hypotenuse of unit length and legs $AC = p$, $BC = q$ (Figure 7.11).

One can see that $\angle ABC$ can be described in the following three ways: $\arcsin p$ (an acute angle whose sine equals p), $\arccos p$ (an acute angle whose cosine equals q), or $\arctan\frac{p}{q}$ (an acute angle whose tangent equals $\frac{p}{q}$). Therefore, for any pair (p, q) of positive numbers, $p^2+q^2=1$, the equalities

$$\arcsin p = \arccos q = \arctan\frac{p}{q} \tag{18}$$

hold true.

Next, if O is the midpoint of AB, then $AO = CO = \frac{1}{2}$.

In order to find $\angle AOC$ one can apply the Law of Cosines to ΔAOC. This results in the equality

$$AC^2 = AO^2 + CO^2 - 2\cdot AO\cdot AO\cdot \cos\angle AOC$$

or

$$p^2 = \frac{1}{2} - \frac{1}{2}\cos(\angle AOC),$$

whence

$$\angle AOC = \arccos(1-2p^2).$$

On the other hand,

$$\angle AOC = \pi - \angle COB = \angle OCB + \angle OBC = 2\cdot\angle ABC.$$

Thus, $\angle ABC = \frac{1}{2}\angle AOC = \frac{1}{2}\arccos(1-2p^2)$ whence

$$\arcsin p = \arccos q = \arctan\frac{p}{q} = \frac{1}{2}\arccos(1-2p^2) \tag{19}$$

Let $p = \frac{1}{\sqrt[4]{3}}$. Then

$$q = \sqrt{1-p^2} = \sqrt{1-\frac{1}{\sqrt{3}}} = \sqrt{1-\frac{\sqrt{3}}{3}} = \sqrt{\frac{3-\sqrt{3}}{3}},$$

$$\frac{p}{q}=\frac{1}{\sqrt[4]{3}}\cdot\sqrt{\frac{3}{3-\sqrt{3}}}=\frac{\sqrt[4]{3^2}}{\sqrt[4]{3}\cdot\sqrt{\sqrt{3}(\sqrt{3}-1)}}$$
$$=\frac{\sqrt[4]{3^2}}{\sqrt[4]{3^2}\cdot\sqrt{\sqrt{3}-1}}=\frac{\sqrt{\sqrt{3}+1}}{\sqrt{(\sqrt{3}-1)(\sqrt{3}+1)}}=\frac{\sqrt{1+\sqrt{3}}}{2},$$

and

$$1-2p^2=1-\frac{2}{\sqrt{3}}=\frac{\sqrt{3}-2}{\sqrt{3}}=\frac{3-2\sqrt{3}}{3}.$$

In that way, when $p=\frac{1}{\sqrt[4]{3}}$, chain (19) turns into chain (17), thereby, formally demonstrating the equivalence of series (10), (12), (14), and (16).

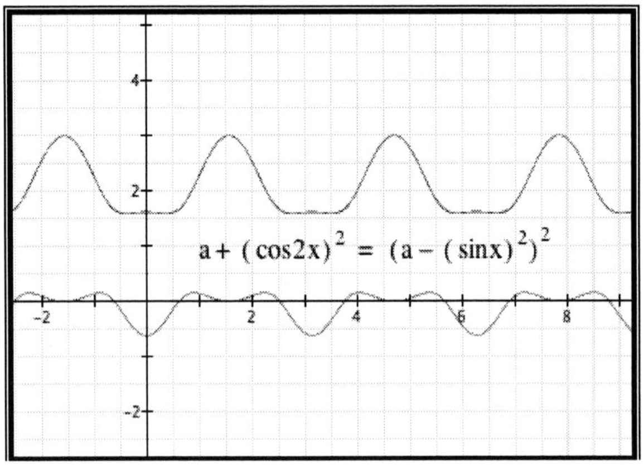

Figure 7.12. Locus of equation (20).

5. Introducing a Parameter in Equation (8)

One may wonder: Is it just a coincidence that both sides of equation (8) include the number 2 as an additive term preceding trigonometric functions? What if one makes this additive term a parameter and parameterize equation (8) to the form

$$a+\cos^2 2x=(a-\sin^2 x)^2 \tag{20}$$

where a is a real parameter? Such a direction in which technology-enhanced discourse can move is consistent with standards-based expectations for secondary students' learning of mathematics that include experience in analyzing mathematical situations described by parametric equations (National Council of Teachers of Mathematics, 2000). Figure 7.12 shows the locus of equation (20) with a custom variable y being used in place of a, just like in

Chapter 2 where algebraic equations with parameters were explored by the *GC*. The locus exhibits several hidden properties of the equation it represents such as the dependence of the number of solutions on parameter *a*, the absence of such solutions not only for sufficiently large *a* (in absolute value), but for relatively small values of *a* as well. That is, equation (20) is very sensitive to the variation of parameter *a*. The locus approach, enabled by the use of the *GC*, allows one to come up with new activities concerning the exploration of trigonometric equations with specified types of solutions. Such activities may include solving a parametric equation for the values of the parameter selected as a result of the analysis of this equation made possible by a computational experiment. Below three such cases will be considered: $a = 3$, $a = 0$, $a = 1$.

5.1. Exploring the Case *a* = 3

One can be asked to solve equation (20) for $a = 3$ using the four methods demonstrated in section 2. Interestingly, unlike the case $a = 2$ associated with equation (8), when $a = 3$ equation (20) has the form

$$3+\cos^2 2x=(3-\sin^2 x)^2$$

Using the four methods demonstrated in section 2 results in the same form of solution

$$x=\frac{\pi}{2}+\pi n \qquad (21)$$

Indeed, the first method leads to the equation $\sin^2 x=1$, the second method leads to the equation $\cos^2 x=0$, and the third method leads to the equation $\cos 2x=-1$. Each of the last three equations has series (21) as the solution. Applying the fourth method leads to the equation $5\cos^4 x+8\sin^2 x\cdot\cos^2 x=0$, the only solution of which is determined by the equation $\cos x=0$ whence (21).

5.2. Exploring the Case *a* = 0

As can be predicted by observing the locus of equation (20) shown in Figure 7.12, a seemingly uncomplicated equation

$$\cos^2 2x=\sin^4 x \qquad (22)$$

requires more work when compared to equation (8). Indeed, the first method leads to the equations

$$\sin^2 x=1 \text{ and } \sin^2 x=\frac{1}{3},$$

the second method leads to the equations

$$\cos^2 x = 0 \text{ and } \cos^2 x = \frac{2}{3},$$

the third method leads to the equations

$$\cos 2x = -1 \text{ and } \cos 2x = \frac{1}{3},$$

and, finally, the fourth method leads to the equations

$$\cos x = 0 \text{ and } \tan^2 x = \frac{1}{2}.$$

We leave to the reader to demonstrate that different methods result in identical series of roots of equation (22).

5.3. Exploring the Case *a* = 1

An interesting exploration is to show, using different methods, that the equation

$$1 + \cos^2 2x = \left(1 - \sin^2 x\right)^2$$

does not have real solutions. This exploration can motivate one to find those values of parameter a for which equation (2) does (or does not) have real solutions. The answer to this query will be given in the next section where equation (8) is generalized to an equation with two parameters.

6. Double Parameterization of Equation (8)

Another way to parameterize equation (8) is to introduce two additive parameters a and b as follows:

$$a + \cos^2 2x = (b - \sin^2 x)^2 \tag{23}$$

In order to address the query posed at the end of the last section, one can use methods described in Chapter 2 and construct a region in the plane of parameters where equation (23) has real solutions. To this end, setting $z = \sin^2 x$ and using formula (9) one can replace equation (23) by the algebraic equation $a + (1-2z)^2 = (b-z)^2$ whence

$$3z^2 + 2(b-2)z + a - b^2 + 1 = 0 \tag{24}$$

Applying quadratic formula to equation (24) and transforming the discriminant as follows

$$(2-b)^2-3(a-b^2+1)=4-4b+b^2-3a+3b^2-3$$
$$=4b^2-4b+1-3a=(2b-1)^2-3a$$

result in

$$z_1=\frac{2-b+\sqrt{(2b-1)^2-3a}}{3} \text{ and } z_2=\frac{2-b-\sqrt{(2b-1)^2-3a}}{3}$$

Consider the first root, z_1. First of all, the discriminant condition requires

$$(2b-1)^2\geq 3a \quad (25)$$

Second, the condition $z\geq 0$ requires

$$\sqrt{(2b-1)^2-3a}\geq b-2 \quad (26)$$

Finally, the condition $z\leq 1$ requires

$$2-b+\sqrt{(2b-1)^2-3a}\leq 3 \quad (27)$$

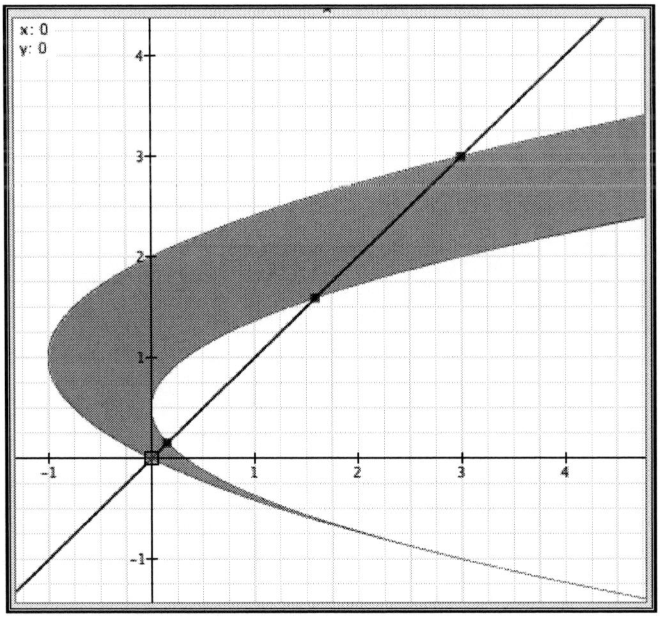

Figure 7.13. The locus of simultaneous inequalities (26) and (27).

Note that inequality (25) is assumed to hold true as the domain of inequalities (26) and (27) and, thereby, can be omitted from graphing separately.

The shaded region in Figure 7.13 shows the locus of simultaneous inequalities (26) and (27) where $a = x$ and $b = y$, constructed by the *GC* as the "graph" of the inequality

$$\sqrt{\sqrt{(2y-1)^2 - 3x} - y + 2\sqrt{1 + y - \sqrt{(2y-1)^2 - 3x}}} > 0 .$$

Using the locus, one can set $b = a$ ($y = x$ in the *GC*) so that the points in common of the locus and the bisector of the first coordinate angle will define those values of parameter a for which equation (20) has real roots. Actually, it is possible to define these values by using the locus of equation (20) and, through cursor pointing to find the intervals $0 < a < 0.157$ and $1.593 < a < 3$. An interesting exercise in solving irrational inequalities is to determine the exact boundaries of the two intervals in the form $(0, \frac{7-\sqrt{33}}{8})$ and $(\frac{7+\sqrt{33}}{8}, 3)$ as solutions to equation (24) when $b = a$, $0 \le z \le 1$. (The points $(\frac{7-\sqrt{33}}{8}, \frac{7-\sqrt{33}}{8})$, $(\frac{7+\sqrt{33}}{8}, \frac{7+\sqrt{33}}{8})$ and (3, 3) are marked on the line $a = b$ in Figure 7.13). In such a way, the shift in focus from solving equation (8) to the study of a two-parametric equation (23) in the context of a computational experiment enables one to appreciate how the use of technology could not only facilitate already challenging curriculum but even push against it.

7. Triple Parameterization of Equation (8)

Consider the equation

$$a + \cos^2 kx = (b - \sin^2 x)^2 \qquad (28)$$

with three real parameters a, b, and k. The locus of equation (28) in the case $a = b = 0$, and $k = 3$ is shown in Figure 7.14. It appears that the equation

$$\cos^2 3x = \sin^4 x \qquad (29)$$

has six roots in the interval (0, π). Could this empirical evidence be confirmed by formal mathematics? Or should solving such an equation be outsourced to software? For example, one can show that setting $z = \sin^2 x$ and using the formula $\cos 3x = 4\cos^3 x - 3\cos x$, the equation $\cos^2 3x = \sin^4 x$ can be reduced to the cubic equation $16z^3 - 23z^2 + 9z - 1 = 0$, the left-hand side of which can be graphed (as the function of z) to demonstrate three intersections with the z-axis in the interval (0, 1). The corresponding graph is shown in Figure 7.15.

Likewise, setting $a = b = 0$ and $k = 5$ in equation (28) yields the equation

$$\cos^2 5x - \sin^4 x = 0 \qquad (30)$$

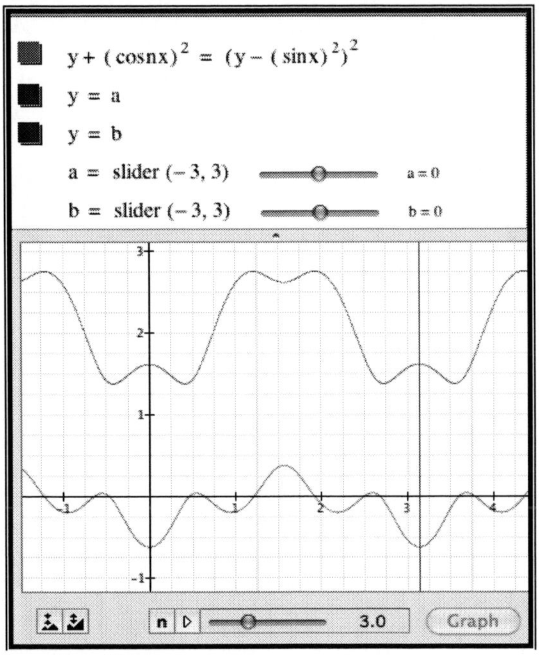

Figure 7.14. The locus of equation (29).

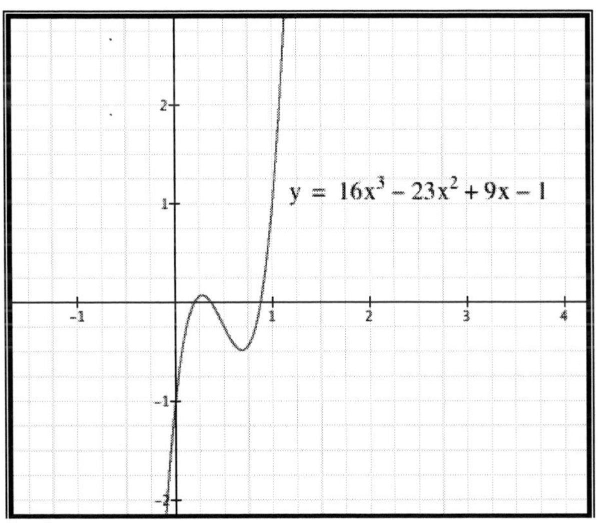

Figure 7.15. The case $k = 3$.

$$\cos(5x)^2 - \sin(x)^4$$

expand(*A1*)

$$256\cos(x)^{10} - 640\cos(x)^8 + 560\cos(x)^6 - 200\cos(x)^4 + 25\cos(x)^2 - \sin(x)^4$$

$$A2 := 256\cdot(1-z)^5 - 640\cdot(1-z)^4 + 560\cdot(1-z)^3 - 200\cdot(1-z)^2 + 25\cdot(1-z) - z^2$$

$$256\,(1-z)^5 - 640\,(1-z)^4 + 560\,(1-z)^3 - 200\,(1-z)^2 + 25 - 25z - z^2$$

simplify(*A2*)

$$1 - 25z + 199z^2 - 560z^3 + 640z^4 - 256z^5$$

Figure 7.16. *Maple*-based transition from equation (30) to equation (31).

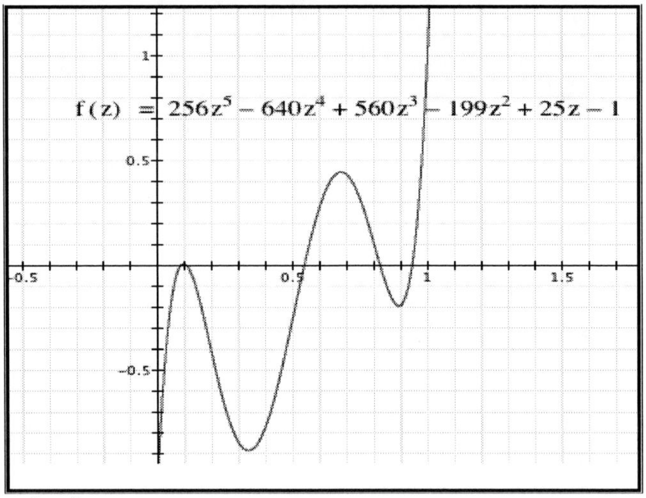

Figure 7.17. The case $k = 5$.

By using *Maple* (Figure 7.16) one can represent the left-hand side of equation (30) as a combination of trigonometric functions of the angle *x*, and then, setting $z = \sin^2 x$, replace equation (30) by the polynomial equation

$$256z^5 - 640z^4 + 560z^3 - 199z^2 + 25z - 1 = 0, \quad (31)$$

the left-hand side of which can be graphed (as the function of *z*) to demonstrate four intersections with the *z*-axis in the interval (0, 1), thereby implying, among other things, the absence of complex roots. As the graph pictured in Figure 7.17 shows, the smallest root of equation (31) is its double root, which provides a local maximum to the corresponding polynomial of degree five.

Finally, considering the case of non-zero values of *a* and *b*, one can establish experimentally (Figure 7.18) the period of the locus of equation (28) for different values of *k*.

8. Exploring the Equation $a\cos t + b\sin t = c$

In this section we will explore the trigonometric equation

$$a\cos t + b\sin t = c \quad (32)$$

where *a*, *b*, *c* are real numbers, $c \neq 0$, and *t* is an angular variable. Without loss of generality, one can consider the equation

$$a\cos t + b\sin t = 1 \quad (33)$$

as equation (32) can be reduced to (33) by substituting *a* for $\frac{a}{c}$ and *b* for $\frac{b}{c}$ when both sides of equation (32) are divided by *c*.

Remark. In Chapter 2, the region in the plane (a, b) where solutions (x, y) of the system of equations $ax + by = 1, x^2 + y^2 = 1$ satisfy the inequality $x > y$ was constructed. For example, as Figure 2.24 (Chapter 2) shows, this region includes the point (2, -2). In terms of equation (33) this means that for any value of t such that $2\cos t - 2\sin t = 1$ the inequality $\cos t > \sin t$ holds true. This fact can be confirmed graphically: in Figure 7.19 the vertical lines represent the values of t for which the *cosine* function always assumes values greater than the *sine* function. In particular, using formula (35) developed below, one can write $\cos(2\arctan\frac{-2+\sqrt{7}}{3}) > \sin(2\arctan\frac{-2+\sqrt{7}}{3})$.

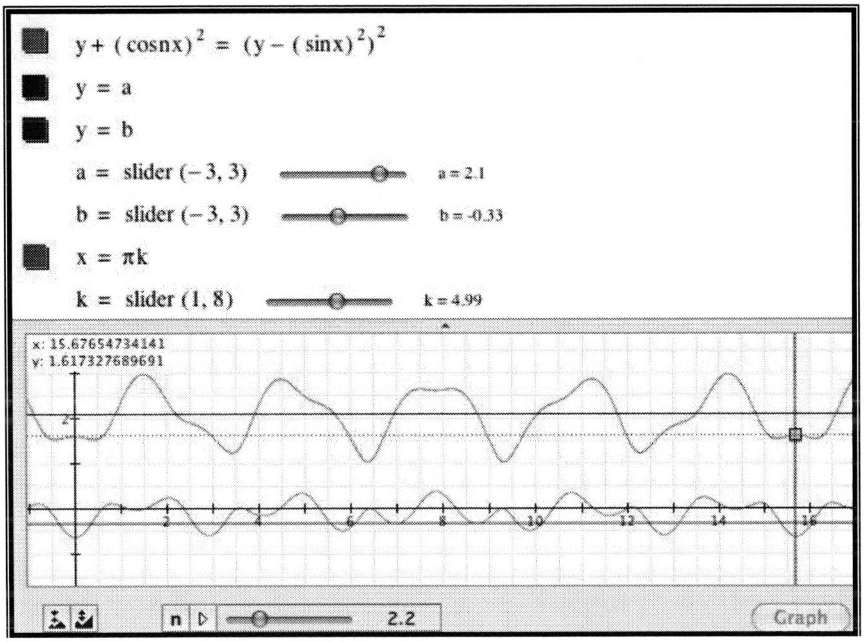

Figure 7.18. Experimentally finding the period of locus of equation (28).

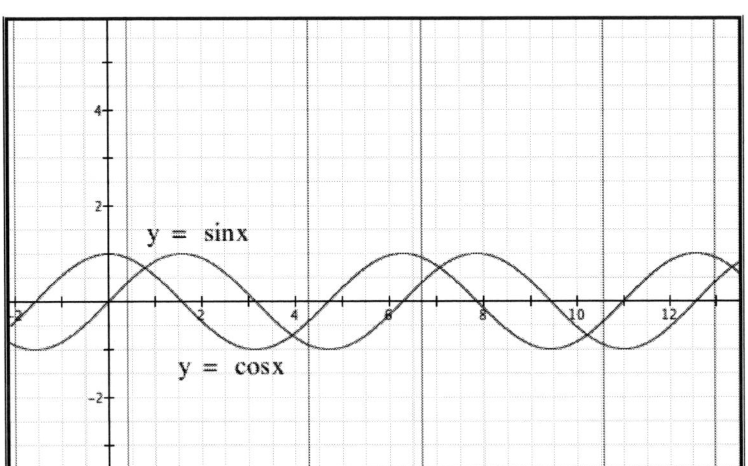

Figure 7.19. When $a = 2$, $b = -2$ in (32), the graph of *cosine* is above the graph of *sine*.

8.1. First Solution to Equation (33)

Equation (33) can be reduced to a homogeneous equation by using the double-angle formulas

$$\cos t = \cos^2\frac{t}{2} - \sin^2\frac{t}{2},\ \sin t = 2\sin\frac{t}{2}\cos\frac{t}{2}$$

and formula (7) and so that

$$a(\cos^2\frac{t}{2} - \sin^2\frac{t}{2}) + 2b\sin\frac{t}{2}\cos\frac{t}{2} = \sin^2\frac{t}{2} + \cos^2\frac{t}{2}.$$

Dividing both sides of the last equality by $\cos^2\frac{t}{2}$ (note: the equation $\cos^2\frac{t}{2} = 0$ does not provide solutions to equation (33)) yields

$$2b\tan\frac{t}{2} + a - a\tan^2\frac{t}{2} = \tan^2\frac{t}{2} + 1$$

whence

$$(a+1)\tan^2\frac{t}{2} - 2b\tan\frac{t}{2} + 1 - a = 0 \qquad (34)$$

Applying the quadratic formula to equation (34) yields

$$\tan\frac{t}{2} = \frac{b \pm \sqrt{a^2+b^2-1}}{1+a}$$

from where, using formula (3), two series can be found

$$t = 2\arctan\frac{b+\sqrt{a^2+b^2-1}}{1+a} + 2\pi n \qquad (35)$$

$$t = 2\arctan\frac{b-\sqrt{a^2+b^2-1}}{1+a} + 2\pi n \qquad (36)$$

8.2. Second Solution to Equation (33)

Another solution is based on the use of an auxiliary angle by representing equation (33) in the form

$$\frac{a}{\sqrt{a^2+b^2}}\cos t+\frac{b}{\sqrt{a^2+b^2}}\sin t=\frac{1}{\sqrt{a^2+b^2}} \qquad (37)$$

Noting that $(\frac{a}{\sqrt{a^2+b^2}})^2+(\frac{b}{\sqrt{a^2+b^2}})^2=1$, one can set $\frac{a}{\sqrt{a^2+b^2}}=\cos\alpha$ and $\frac{b}{\sqrt{a^2+b^2}}=\sin\alpha$ so that $\alpha=\arctan\frac{b}{a}$ and equation (37) can be written as

$$\cos t\cdot\cos\alpha+\sin t\cdot\sin\alpha=\frac{1}{\sqrt{a^2+b^2}} \text{ or } \cos(t-\alpha)=\frac{1}{\sqrt{a^2+b^2}}$$

whence, using formula (2),

$$t-\alpha=\pm\arccos\frac{1}{\sqrt{a^2+b^2}}+2\pi n$$

or

$$t=\arctan\frac{b}{a}+\arccos\frac{1}{\sqrt{a^2+b^2}}+2\pi n \qquad (38)$$

and

$$t=\arctan\frac{b}{a}-\arccos\frac{1}{\sqrt{a^2+b^2}}+2\pi n \qquad (39)$$

8.3. Demonstration of the Case *a* = 2, *b* = -2

One can use the *GC* to demonstrate that in the case $a = 2$, $b = -2$, that is, for the equation $2\cos t-2\sin t=1$, the series of roots given by formulas (35)-(36) and (38)-(39) are identical. More specifically, the following identities hold true:

$$\arctan(-1)+\arccos\frac{1}{\sqrt{8}}=2\arctan\frac{-2+\sqrt{7}}{3} \qquad (40)$$

and

$$\arctan(-1)-\arccos\frac{1}{\sqrt{8}}=2\arctan\frac{-2-\sqrt{7}}{3} \qquad (41)$$

Identities (40) and (41) can first be verified graphically, as shown in Figure 7.20, and then proved formally by using the following combination of argument and computation. Consider identity (40). One can show that both sides of (40) belong to the interval $(0, \frac{\pi}{2})$ where the function $f(\alpha) = \tan\alpha$ monotonically increases so that the equality $f(\alpha) = f(\beta)$ implies $\alpha = \beta$ provided $\alpha, \beta \in (0, \frac{\pi}{2})$.

To this end note that $0 < \frac{\sqrt{7}-2}{3} < 1$ and, therefore,

$$0 < \arctan\frac{\sqrt{7}-2}{3} < \frac{\pi}{4}$$

whence

$$0 < 2\arctan\frac{\sqrt{7}-2}{3} < \frac{\pi}{2}.$$

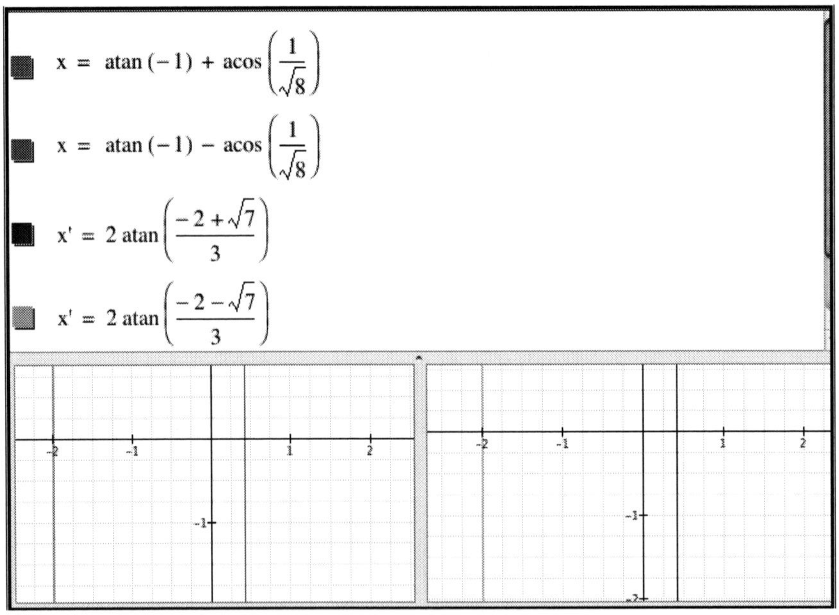

Figure 7.20. Computer demonstration of identities (40) and (41).

Next, the inequalities $0 < \frac{1}{\sqrt{8}} < \frac{1}{\sqrt{2}}$ imply

$$\frac{\pi}{4} = \arccos\frac{1}{\sqrt{2}} < \arccos\frac{1}{\sqrt{8}} < \arccos 0 = \frac{\pi}{2}$$

whence

$$0 < -\frac{\pi}{4} + \arccos\frac{1}{\sqrt{8}} < \frac{\pi}{4}.$$

Setting $\alpha = -\frac{\pi}{4} + \arccos\frac{1}{\sqrt{8}}$ and $\beta = 2\arctan\frac{\sqrt{7}-2}{3}$ one has the required inclusion $\alpha, \beta \in (0, \frac{\pi}{2})$. Using the formula $\tan(\alpha_1 + \alpha_2) = \frac{\tan\alpha_1 + \tan\alpha_2}{1 - \tan\alpha_1 \cdot \tan\alpha_2}$ and noting that $p = \sqrt{\frac{7}{8}}$ and $q = \frac{1}{\sqrt{8}}$ in (18) yields the equality $\arccos\frac{1}{\sqrt{8}} = \arctan\sqrt{7}$, one can proceed with computations as follows

$$\tan\alpha = \tan(-\frac{\pi}{4} + \arccos\frac{1}{\sqrt{8}}) = \frac{\tan(-\frac{\pi}{4}) + \tan(\arccos\frac{1}{\sqrt{8}})}{1 - \tan(-\frac{\pi}{4}) \cdot \tan(\arccos\frac{1}{\sqrt{8}})} = \frac{-1+\sqrt{7}}{1+\sqrt{7}}$$

and

$$\tan\beta = \frac{2\tan(\arctan\frac{\sqrt{7}-2}{3})}{1 - [\tan(\arctan\frac{\sqrt{7}-2}{3})]^2} = \frac{6(\sqrt{7}-2)}{9-(\sqrt{7}-2)^2} = \frac{6(\sqrt{7}-2)}{(1+\sqrt{7})(5-\sqrt{7})}.$$

To complete the demonstration of identity (40) it remains to be shown that $\frac{-1+\sqrt{7}}{1+\sqrt{7}} = \frac{6(\sqrt{7}-2)}{(1+\sqrt{7})(5-\sqrt{7})}$.

Indeed, $(-1+\sqrt{7})(5-\sqrt{7}) = -5-7+6\sqrt{7} = 6(\sqrt{7}-2)$.

The formal proof of identity (41) is left to the reader.

8.4. Third Solution to Equation (33).

The remark made at the beginning of section 8 implied that equation (33) can be reduced to a system of two simultaneous equations

$$ax + by = 1, x^2 + y^2 = 1 \tag{42}$$

in variables x and y where

$$x = \cos t, \quad y = \sin t \tag{43}$$

As shown in Chapter 2 (formulas (27)-(28)), system (42) has the following solutions

$$(x, y) = (\frac{a + b\sqrt{a^2 + b^2 - 1}}{a^2 + b^2}, \frac{b - a\sqrt{a^2 + b^2 - 1}}{a^2 + b^2})$$

$$(x, y) = (\frac{a - b\sqrt{a^2 + b^2 - 1}}{a^2 + b^2}, \frac{b + a\sqrt{a^2 + b^2 - 1}}{a^2 + b^2})$$

where $a^2 + b^2 \geq 1$. In order to use the above values of x and y in transition to the values of the angular variable t simultaneously satisfying relations (43), three cases have to be considered.

1. Let $y \in [-1, 1], x \in [0, 1]$. Then, it follows from Figure 7.2 and the 2π-periodicity of $\cos t$ and $\sin t$ that

   $$t = \arcsin y + 2\pi n \tag{44}$$
2. Let $x \in [-1, 1], y \in [0, 1]$. Then, it follows from Figure 7.4 and the 2π-periodicity of $\cos t$ and $\sin t$ that

   $$t = \arccos x + 2\pi n \tag{45}$$

3. Let $x \in [-1, 0], y \in [-1, 0]$. Then it follows from Figures 7.2 and 7.4 that the main value of t belongs to the third quadrant and cannot be expressed either through $\arccos x$ or $\arcsin y$, as these two values, by definition, do not belong to the third quadrant. In this case, we have

   $$t = -\pi + \arccos|x| + 2\pi n \tag{46}$$

or, alternatively, $t = -\pi + \arcsin|y| + 2\pi n$.

Therefore, given the pair (a, b), one has to decide which of the three cases each of the two solutions to system of equations (42) satisfies, and then use formulas (44)-(46) to represent the series of roots of equation (33). For example, when (a, b) = (2, -2), we have $(x_1, y_1) = (\frac{1 - \sqrt{7}}{4}, \frac{-1 - \sqrt{7}}{4})$—case 3, which yields $t = -\pi + \arccos\frac{\sqrt{7} - 1}{4} + 2\pi n$. At the same time, $(x_2, y_2) = (\frac{1 + \sqrt{7}}{4}, \frac{-1 + \sqrt{7}}{4})$—case 1 (or case 2), which yields

$t = \arcsin\frac{\sqrt{7} - 1}{4} + 2\pi n$ (alternatively, $t = \arcsin\frac{\sqrt{7} + 1}{4} + 2\pi n$).

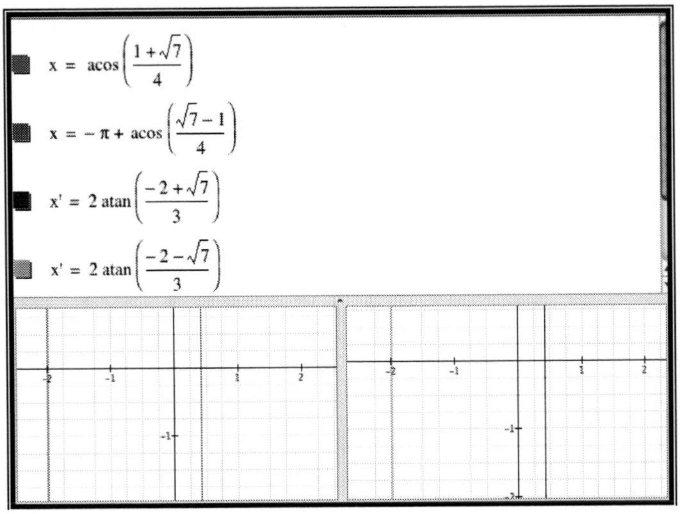

Figure 7.21. Computer demonstration of identities (47) and (48).

Figure 7.21 demonstrates how in the case $(a, b) = (2, -2)$ the series of roots of equation (33) obtained through the first and the third methods generate same points on the number line. A formal demonstration of the identities

$$\arccos\frac{\sqrt{7}+1}{4} = 2\arctan\frac{\sqrt{7}-2}{3} \tag{47}$$

and

$$-\pi + \arccos\frac{\sqrt{7}-1}{4} = -2\arctan\frac{2+\sqrt{7}}{3} \tag{48}$$

is left to the reader.

9. ACTIVITY SET

1. Demonstrate graphically and analytically that the equation

$$\sin^5 x + 10\sin^4 x + 36\sin^3 x + 56\sin^2 x + 35\sin x + 6 = 0$$

has the following two series of roots: $x = -\frac{\pi}{2} + 2\pi n$ and $x = (-1)^n \arcsin(-2+\sqrt{3}) + \pi n$.

2. Demonstrate graphically and analytically (using formulas (1)-(3) in finding x) that the equations $\sin^2 x = \frac{1}{3}$, $\cos^2 x = \frac{2}{3}$, $\cos 2x = \frac{1}{3}$, and $\tan^2 x = \frac{1}{2}$ have identical series of roots.
3. Prove the identity

$$\frac{\pi}{4} + \arcsin\frac{\sqrt{14}}{4} = 2\arctan\frac{2+\sqrt{7}}{3}.$$

4. Prove the identity

$$\arccos\frac{\sqrt{7}+1}{4} = 2\arctan\frac{\sqrt{7}-2}{3}.$$

5. Prove the identity

$$\arccos\frac{\sqrt{7}-1}{4} + 2\arctan\frac{2+\sqrt{7}}{3} = \pi.$$

Chapter 8

DEVELOPING MODELS FOR COMPUTATIONAL PROBLEM SOLVING

Most mathematical concepts or generalizations can be effectively introduced using a problem situation.

— National Council of Teachers of Mathematics (2000, p. 335)

1. INTRODUCTION

One of the main directions of the reform of mathematics curriculum and pedagogy put forth by the National Council of Teachers of Mathematics (1989, 2000) is teaching through modeling and applications. Making mathematics relevant to the outside world, solving problems in context, learning to use mathematical concepts as models and tools, utilizing and connecting these concepts in computing application—these are some of the basic ideas that underpin new vision of the mathematics classroom at the pre-college level. In turn, recommendations for teacher preparation include the need to provide technology-enhanced training in modeling-oriented pedagogy in a capstone course (Conference Board of the Mathematical Sciences, 2001). Teachers with experience of using mathematical concepts as tools and models in computing applications are more likely to have a success in introducing ideas of the reform to their students than those with the traditional experience in teaching to the production of correct answers.

As has been continuously demonstrated throughout the book, the use of a spreadsheet has great potential to enrich mathematical modeling pedagogy. In particular, using the software, teachers can "explore fundamental properties of arithmetic, geometric, and more general sequences and series" (ibid, p. 128) appropriately selected to enable sophisticated computations. In doing so, teachers can come across many computationally driven problematic situations with no apparent relevance to original contextual inquiries for which their models were designed. To solve new problems, new computational environments might have to be developed that, in turn, prompt new inquiries and stimulate search of new problem-solving strategies.

In this chapter, once again, we use polygonal numbers as principal mathematical concepts. In the very beginning of the book triangular and square numbers were mentioned in the context of summation of consecutive counting and odd numbers, respectively. This context can be extended to introduce polygonal numbers as partial sums of more general arithmetic series. It is through such generalization that one comes to understand how these new concepts can be used as tools applicable to situational problem-solving extensions.

In problematic situations discussed in this chapter two ways of formulating their extensions will be emphasized. One such way is to extend problematic situation by altering the corresponding context. In doing so, after a mathematical model of contextual inquiry has been developed, one goes back and changes the inquiry, develops a relevant model involving a number of parameters, and refines corresponding problem-solving tools. Apparently, this approach does not require one's grasp of generalized meanings of parameters involved in the construction of the original model. As a result, mathematical concepts that emerge through the multiple implementation of this approach may not be formally connected to each other.

Another way of extending a problematic situation is to change parameters of a model within the model itself and formulate a new contextual inquiry through interpreting this change in terms of context. This change, however, does require one's conceptualization of parameters involved in the original model and the resulting mathematical concepts are likely to be connected to each other through the layers of consecutive generalizations. In other words, whereas the first approach to posing new problems describes situation when model is dependent on contextual inquiry (but not vice versa), the second approach describes situation when contextual inquiry results from the meaningful change (or extension) of parameters of a model. The diagrams of Figures 8.1 and 8.2 illustrate the difference in the two approaches to formulating problematic situations and developing mathematical models to be used in computing applications.

Figure 8.1. Model does not affect contextual inquiry.

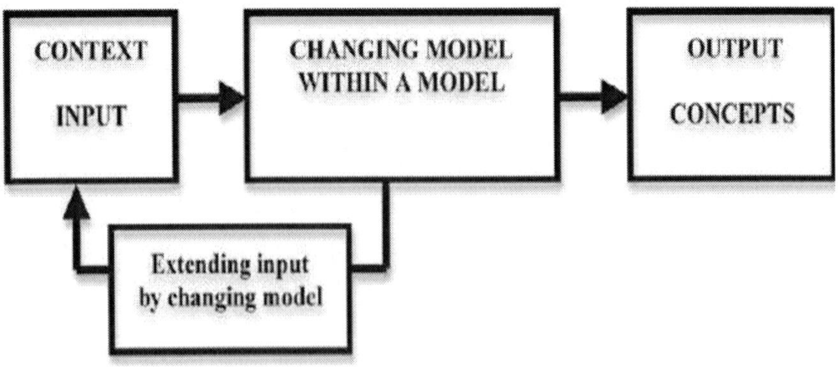

Figure 8.2. Change of model affects contextual inquiry.

2. SETTING A CONTEXT AND INTRODUCING MODELING TOOLS

In what follows, computational problem solving will be considered in a whimsical context. It should be noted that one's perception of what is whimsical and what is not depends on experience. Indeed, many of today's real-life situations seemed like fictions in the recent past. To begin, consider the following context.

A young architect wanted to extend the concept used at the construction of the Pentagon and proposed using pentagonal numbers 1, 5, 12, 22, 35, 51, ... (he learned about these numbers in a collegiate number theory course) for the construction of what he called a pentagonal hotel, the blueprint of which is pictured in Figure 8.3. The hotel was made up of a number of buildings adjacent to each other. Each building had one, four, seven, ten, thirteen, or sixteen stories with one room on each storey...

Below different problematic situations (contextual inquiries) that stem from this context will be created and then resolved through mathematical modeling activities leading to a spreadsheet-based computerization of the constructed models. Finally, two types of problematic situations will be considered: a numerical one for which a solution will be suggested and its generalization for which a model based on different mathematical tools will be developed. In referring to context, the following terminology will be used. Any one-cell unit will be referred to as *room*. A number used to label a cell will be referred to as *room number*. A combination of one or more vertically arranged rooms will be referred to as *building*. A combination of different buildings adjacent to each other will be referred to as *block*. Finally, a combination of several blocks will be referred to as *hotel*. In such a way, there are 204 rooms, twenty-four buildings, and four blocks in the blueprint of the 16-storied hotel pictured in Figure 8.3. It is such a hotel that the architect referred to as the pentagonal hotel of rank six.

One can see that the sequence of the number of rooms (alternatively, stories) in the buildings comprising the first block, forms the arithmetic sequence

$$1, 4, 7, 10, 13, 16, \ldots \qquad (1)$$

with the difference three. In turn, the partial sums of sequence (1) form a new number sequence

$$1, 1+4, 1+4+7, 1+4+7+10, 1+4+7+10+13,$$
$$1+4+7+10+13+16, \ldots$$

One can interpret the above sums in general terms by associating them with the sequence of evolving pentagons (Figure 8.4), which develop from a single dot in such a way that each side of a new pentagon gets a new dot. Counting the dots associated with each pentagon yields the number sequence

$$1, 5, 12, 22, \ldots \qquad (2)$$

called pentagonal numbers.

In constructing mathematical models associated with the context of pentagonal hotels, the following five spreadsheet-based functions will be used. The function MOD(x, y) returns the remainder when x is divided by y. The function INT(x) returns the greatest integer smaller or equal to x. The function CEILING(x, y) returns the smallest number greater than x and divisible by y. The function FLOOR(x, y) returns the largest number smaller that x and divisible by y. Finally, the function ROUND(x, 0) returns the integer closest (or equal) to x.

					51						102						153						204
					50						101						152						203
					49						100						151						202
				35	48					86	99					137	150					188	201
				34	47					85	98					136	149					187	200
				33	46					84	97					135	148					186	199
			22	32	45				73	83	96				124	134	147				175	185	198
			21	31	44				72	82	95				123	133	146				174	184	197
			20	30	43				71	81	94				122	132	145				173	183	196
		12	19	29	42			63	70	80	93			114	121	131	144			165	172	182	195
		11	18	28	41			62	69	79	92			113	120	130	143			164	171	181	194
		10	17	27	40			61	68	78	91			112	119	129	142			163	170	180	193
	5	9	16	26	39		56	60	67	77	90		107	111	118	128	141		158	162	169	179	192
	4	8	15	25	38		55	59	66	76	89		106	110	117	127	140		157	161	168	178	191
	3	7	14	24	37		54	58	65	75	88		105	109	116	126	139		156	160	167	177	190
1	2	6	13	23	36	52	53	57	64	74	87	103	104	108	115	125	138	154	155	159	166	176	189

Figure 8.3. A blueprint of four adjacent pentagonal hotels of rank six.

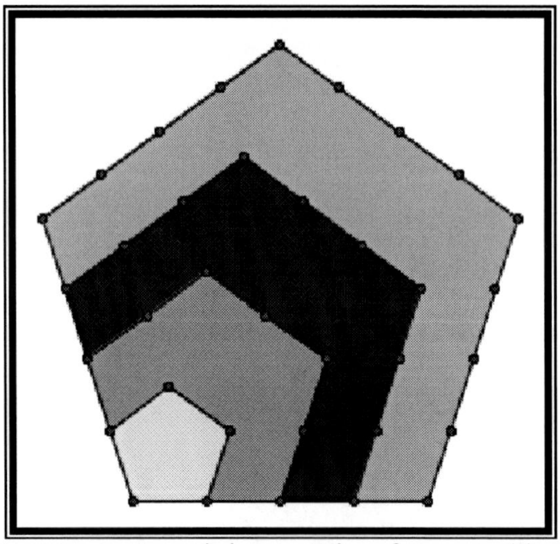

Figure 8.4. Evolving pentagons as a geometric interpretation of sequence (2).

3. MODELING VS. COUNTING

Observing the blueprint pictured in Figure 8.3, one can see that in the first block room numbers located at the top storey of each building are pentagonal numbers. Consequently, in other blocks such room numbers are pentagonal numbers modulo 51—the pentagonal number of rank six (alternatively, the largest room number in the first block). One may wonder if there is any relationship between building number and the number of rooms in this building? With this in mind, the following simple problematic situation (PS) can be formulated:

PS 1. How many rooms are in the 20th building of the pentagonal hotel of rank six?

Solution. Although one can easily answer the question by counting 20 consecutive buildings and then counting the number of rooms in the 20th building as shown in the blueprint of Figure 8.3, such elementary approach, not requiring any mathematical skill beyond counting, would be limited to the use of the diagram. To open window to the use of mathematics, one can observe that the number of rooms in each building of the first block is repeated in the corresponding buildings in all other blocks. One can say that the number of rooms in a building is a periodic function of the building number. More specifically, all buildings whose numbers have the same remainder when divided by six—the hotel's rank—have the same number of rooms. In particular, this means that building 20 has the same number of rooms as building 2 because 20 gives remainder 2 when divided by 6. In other words, the numbers 20 and 2 are congruent modulo 6. Using the notation MOD(x, y), one has the equality MOD(20, 6) = 2. Note that the only exception to this rule is the zero remainder. For example, the equality MOD(x, 6) = 0 implies that the 6th and x-th buildings have the same number of rooms; that is, all buildings numerated by multiples of six have the same number of rooms (alternatively, stories) as building 6.

The second step, once again, can be mathematized. That is, deciding the number of rooms in building 2 can be made independent on the availability of the blueprint. Indeed, as the largest room number in each building of the first block is a pentagonal number (this follows from the rule according to which the hotel was designed), the number of rooms in each such building can be found as the difference between two consecutive pentagonal numbers. In turn, the differences between two consecutive pentagonal numbers form an arithmetic sequence $d(n)$ with the first term one and difference three (for convenience, we define pentagonal number of rank zero as zero). Such a sequence has the form $d(n) = 3n - 2$ where $n = 1, 2, \ldots,$ 6—building numbers in the first block. Note that the number of rooms in each building coincides with the number of stories in the building; in other words, a hotel with n buildings in a block has $3n - 2$ stories. As $d(2) = 4$, there are four rooms in the 20th building of the hotel. Such formalization is required when one develops a computerized model for deciding the number of rooms in a building, given its number. With this in mind, consider

PS 2. How many rooms are in building B of the pentagonal hotel of rank n?

Developing a model. First, we have to find building number in the first block that is equal to the remainder of B when divided by n. If MOD(B, n) > 0, then MOD(B, n) gives such a number; otherwise, the building number is equal to n. The second step is to substitute n for MOD(B, n) > 0 (otherwise, leave n without change) in the formula $d(n) = 3n - 2$. This gives d(MOD(B, n)) = 3·MOD(B, n) – 2. In other words,

if MOD(B, n) > 0 there are 3·MOD(B, n) – 2 rooms in building B;

if MOD(B, n) = 0 building B has $3n - 2$ rooms.

The described model is pictured in the diagram of Figure 8.5. When $B = 408$ and $n = 54$ one has 3·MOD(408, 54) = 3·30 – 2 = 88 —the number of rooms in building 408 of the pentagonal hotel with 54 buildings in a block. At the same time, when $B = 408$ and $n = 17$ one has MOD(408, 17) = 0 and thus there are 49 rooms in building 408. Computations of that kind can be done by a spreadsheet. Figure 8.6 shows the spreadsheet with cells *A3* and *G3* serving as input (*B* and *n*, respectively) and cell *I3* serving as output (the number of rooms).

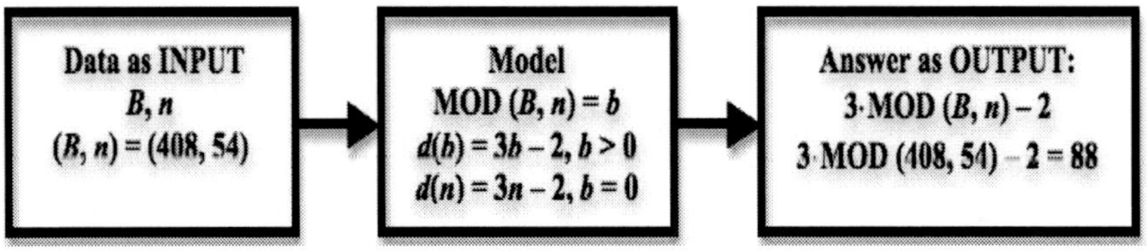

Figure 8.5. Model development of PS 2.

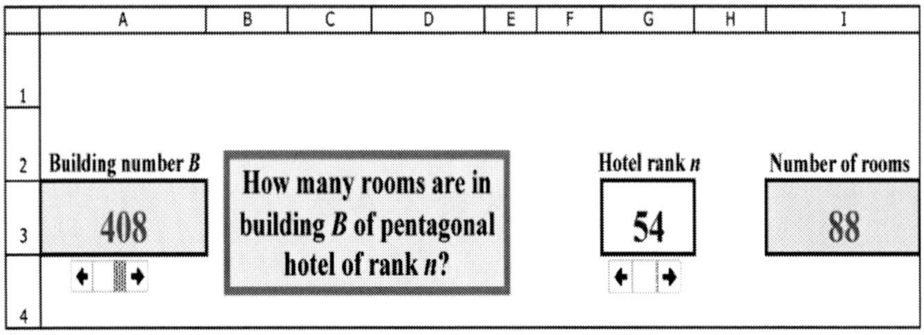

Figure 8.6. Spreadsheet modeling of PS 2.

Remark 1. The concept of congruence serves as a mathematical model for Problems 1 and 2 . Numbers *a* and *b* are referred to as being congruent modulo *m* if they have the same remainder when divided by *m*. This concept brings about the associated notion of remainder described by the function MOD. The importance of this function as a tool in developing various mathematical models will be continuously demonstrated throughout the chapter.

4. Polygonal Numbers Revisited

Extending the above inquiry into the number of rooms, given building number, to other types of hotels can motivate an alternative introduction of polygonal numbers mentioned earlier in Chapter 6. Like pentagonal numbers (2), polygonal numbers, in general, can be introduced also as partial sums of an arithmetic series depending on two parameters. To this end, consider the first *n* terms of the arithmetic series with the first term one and difference $m - 2$:

$$1, m-1, 2m-3, 3m-5, \ldots, (n-1)m-(2n-3) \tag{3}$$

The sum of these numbers, i.e., the *n*-th partial sum of the arithmetic series is exactly $P(m, n)$—the polygonal number of side *m* and rank *n*. In order to find this sum, one has to increase *n*-fold the average of the first and the last terms of sequence (3). In doing so, one can write

$$P(m,n) = \frac{1+(n-1)m-2n+3}{2}n$$

or

$$P(m,n) = \frac{n(n-1)}{2}(m-2)+n \qquad (4)$$

Formula (4), a closed formula for the polygonal number of side *m* and rank *n* (alternatively, the *m*-gonal number of rank *n*), coincides with formula (18) of Chapter 6 developed from the recursive definition of polygonal numbers.

In particular, when *m* = 5 the expression $P(5,n) = \frac{3}{2}n(n-1)+n = \frac{n(3n-1)}{2}$ is the pentagonal number of rank *n*. One can check to see that the first difference $\Delta_1 P(5,n) = 3n-2$. Indeed,

$$\Delta_1 P(5,n) = P(5,n) - P(5,n-1) = \frac{n(3n-1)}{2} - \frac{(n-1)(3n-4)}{2} = 3n-2\,.$$

In general, the first difference for the sequence of polygonal numbers

$$\Delta_1 P(m,n) = P(m,n) - P(m,n-1) = \frac{n(n-1)}{2}(m-2)$$

$$+n - \frac{(n-1)(n-2)}{2}(m-2) - (n-1)$$

$$= \frac{(n-1)}{2}(m-2)(n-n+2)+1 = (m-2)n-(m-3)$$

and, thereby, $\Delta_1 P(m,n)$ is the *n*-th term of the arithmetic sequence

$$d(m, n) = (m-2)n - (m-3) \qquad (5)$$

Contextually, formula (5) gives the number of stories in the *m*-gonal hotel of rank *n*. For example, when *m* = 5 it follows from formula (5) that *d*(5, *n*) = 3*n* – 2 – the number of stories in the pentagonal hotel of rank *n*.

5. Change of Model Affects Context

The introduction of polygonal numbers *P*(*m*, *n*) enables one to consider a more general model (or models) and, as a result, to seek a new context to which the extended model can be applied. This situation is described in the diagram of Figure 8.2: the change of model within a model affects context/input. As an illustration, consider

PS 3. How many rooms are in building *B* of the polygonal hotel of side *m* and rank *n*?

Developing a model. As shown in the diagram of Figure 8.7, ascribing meaning to formula (5) enables one to change a model within a model and, thereby, accommodate a more general input (in comparison with Figure 8.5). That is, the first step is to map building *B* to the 1st block by calculating MOD(*B*, *n*). The second step is to use formula (5) and, if MOD(*B*, *n*) > 0, find *d*(*m*, MOD(*B*, *n*))—the number of rooms sought. When MOD(*B*, *n*) = 0, buildings *B* and *n* have the same number of rooms; that is, formula (5) gives the number of rooms in building *B*.

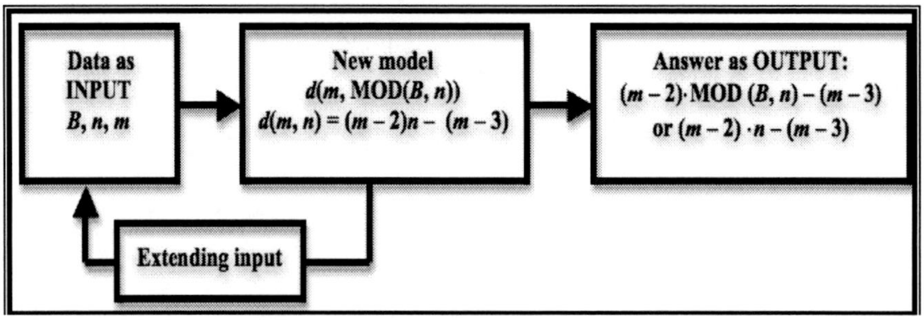

Figure 8.7. The change of model within a model (modeling PS 3).

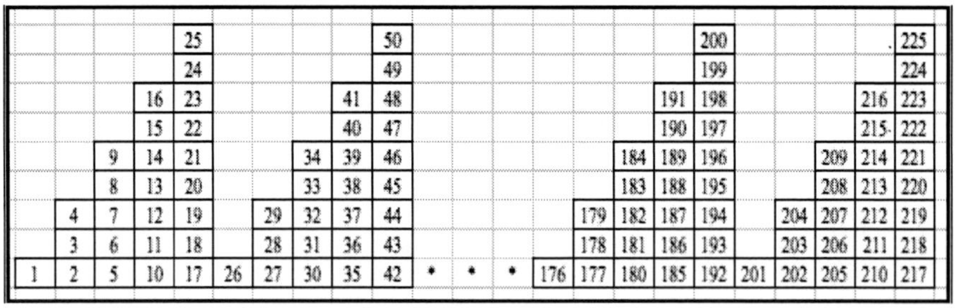

Figure 8.8. Fragments of the 4-gonal hotel of rank 5.

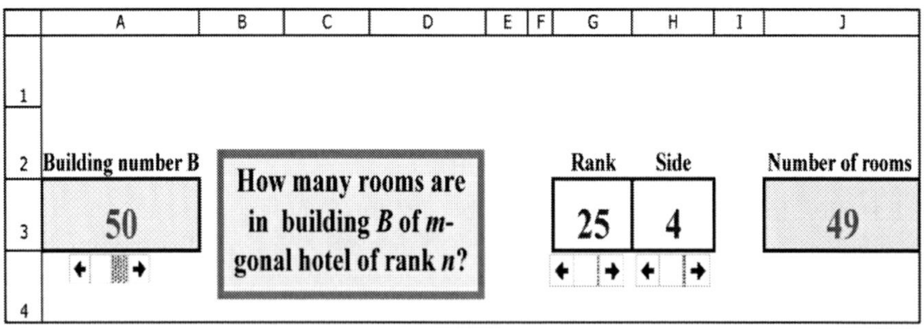

Figure 8.9. Spreadsheet modeling of PS3.

For example, as the blueprint pictured in Figure 8.8 confirms, when $B = 43$, $m = 4$, and $n = 5$, one has MOD(43, 5) = 3, therefore, $d(4, 3) = (4 - 2)\cdot 3 - (4 - 3) = 5$—the number of rooms in building 43. Once again, a spreadsheet can be used to computerize PS 3 (Figure 8.9) to accommodate any triple (*B*, *n*, *m*).

6. Change of Context Requires New Model

Consider the case described by the diagram of Figure 8.1 when a new model results from the change of context rather than from the variation of an old model.

PS 4. To which building of the pentagonal hotel of rank six does room 120 belong?

Solution. Note that the sequences of room numbers that belong to the same storey represent equivalent classes of numbers with identical residues modulo 51—the pentagonal number of rank six. Repeatedly subtracting 51 from 120 until a number smaller than 51 is reached, yields 18—the room number in the first block congruent to 120 modulo 51. In other words, $18 = MOD(120, 51)$.

The next step is to find the rank of the smallest pentagonal number, $\frac{n(3n-1)}{2}$, greater than (or, in general, equal to) 18. This can be done by rounding up to the nearest integer the smallest positive solution of the inequality $\frac{n(3n-1)}{2} \geq 18$ which is equivalent to $3n^2 - n - 36 \geq 0$. To this end, by using the quadratic formula, one has to find $CEILING(\frac{1+\sqrt{1+432}}{6}, 1) = 4$—the building number (in the first block) to which room 18 belongs. Finally, one has to map building 4 back to the block where room 120 belongs and then find building number congruent to 4 modulo 6 as follows:

$$4 + 6 \cdot INT(120/51) = 4 + 6 \cdot 2 = 16.$$

Therefore, room 120 belongs to building 16.

At a more general level, one can formulate

PS 5. To which building of the *m*-gonal hotel of rank *n* does room *r* belong?

Developing a model. Using formula (4), one can find that the total number of rooms in the first block is equal to $\frac{n(n-1)}{2}(m-2)+n$, and then map room *r* to the first block by calculating

$$r_1 = MOD(r, \frac{n(n-1)}{2}(m-2)+n)$$

The next step is to find the rank of the smallest *m*-gonal number greater than or equal to r_1. To this end, one has to solve the inequality

$$\frac{n(n-1)}{2}(m-2)+n \geq r_1,$$

which can be simplified to the form

$$(m-2)n^2+(4-m)n-2r_1 \geq 0.$$

The last inequality turns into equality when

$$n=\frac{m-4\pm\sqrt{(m-4)^2+8(m-2)r_1}}{2(m-2)}.$$

Considering the positive value of n, one has to find

$$CEILING(\frac{m-4+\sqrt{(m-4)^2+8(m-2)r_1}}{2(m-2)},1) \quad (6)$$

the value of which gives the building number to which room r_1 belongs.

Finally, one has to find the building number to which room r belongs by calculating

$$INT(\frac{r}{0.5n(n-1)(m-2)+n})$$

and then the sum

$$CEILING(\frac{m-4+\sqrt{(m-4)^2+8(m-2)r_1}}{2(m-2)},1)$$
$$+n\cdot INT(\frac{r}{0.5n(n-1)(m-2)+n}),$$

which represents the building number sought. This completes the development of model for PS 5.

Example 1. When $m = 5$, $n = 6$ and $r = 120$, the last sum yields

$$CEILING(\frac{1+\sqrt{1+24\cdot MOD(193/51)}}{6},1)$$
$$+6\cdot INT(\frac{193}{3\cdot 5\cdot 3+6})=4+6\cdot 2=16$$

These calculations, confirmed by the spreadsheet pictured in Figure 8.10, show how the general model designed in the course of resolving PS 5 can be verified over the specific case explored in PS 4.

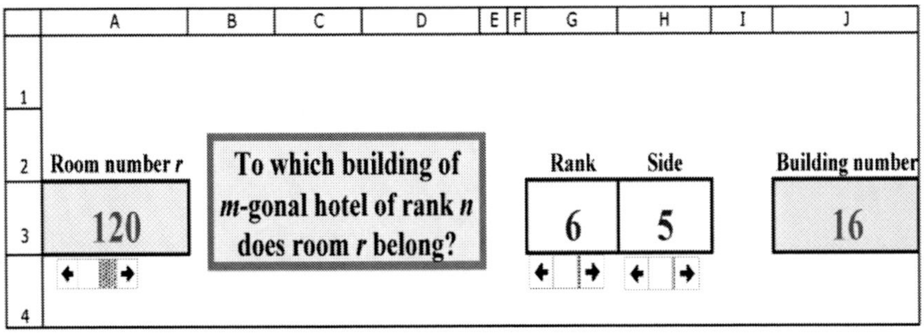

Figure 8.10. Spreadsheet modeling of PS 5.

Now, once again, the change of context will be shown as a way of formulating new inquiries and developing new models. In doing so, one can use certain elements of already constructed models as parts of these new models. As an illustration, consider

PS 6. To which storey of the pentagonal hotel of rank 6 does room 193 belong?

Solution. One can use a part of the model described in PS 5 to find (within the first block) room 40, which, due to the equality $MOD(193,51)=40$ is located on the same storey as room 193. In order to proceed further this time, one has to find the largest pentagonal number smaller than (or equal to) 40 by solving the inequality $\frac{n(3n-1)}{2}\le 40$, which is equivalent to $3n^2-n-80\le 0$, and then rounding down to the nearest integer the largest solution of this inequality. To this end, using the quadratic formula, one can calculate

$$FLOOR(\frac{1+\sqrt{1+960}}{6},1)=5.$$

and then find

$$p_5=\frac{n(3n-1)}{2}\Big|_{n=5}=35.$$

Finally, subtracting 35 from 40 yields 5—the storey number to which room 193 belongs.

Now, consider the general case of the *m*-gonal hotel of rank *n*.

PS 7. To which storey of the *m*-gonal hotel of rank *n* does room *r* belong?

Developing a model. First, we have to solve the following auxiliary problem: Find $R(Q)$—the rank of the largest *m*-gonal number smaller than Q. To this end, the inequality

$$\frac{n(n-1)}{2}(m-2)+n<Q$$

has to be transformed to the standard form of a quadratic inequality (in variable *n*)

$$(m-2)n^2-n(m-4)-2Q<0$$

whence

$$n < \frac{m-4+\sqrt{(m-4)^2+8(m-2)Q}}{2(m-2)}.$$

One can check to see that when $m = Q$ the last inequality turns into $n < 2$ implying the equality $R(Q) = 1$. Therefore, when

$$INT[\frac{m-4+\sqrt{(m-4)^2+8(m-2)Q}}{2(m-2)}]$$
$$\neq \frac{m-4+\sqrt{(m-4)^2+8(m-2)Q}}{2(m-2)}$$

(7)

we have

$$R(Q) = INT[\frac{m-4+\sqrt{(m-4)^2+8(m-2)Q}}{2(m-2)}],$$

otherwise

$$R(Q) = \frac{m-4+\sqrt{(m-4)^2+8(m-2)Q}}{2(m-2)} - 1$$
$$= \frac{m-4+\sqrt{(m-4)^2+8(m-2)Q}-2m+4}{2(m-2)}$$
$$= \frac{-m+\sqrt{(m-4)^2+8(m-2)Q}}{2(m-2)}.$$

Now, one has to find $P(m, R(Q))$ where $Q = MOD(r, P(m,n))$. Note that the case $MOD(r, P(m,n)) = 0$ implies that room r belongs to the top storey the number of which can be found through formula (5); that is, the storey number sought is equal to $(m-2)n-(m-3)$. Finally, assuming that $MOD(r, P(m,n)) \neq 0$, one has to calculate the difference $Q - P(m, R(Q))$ which represents the storey number to which room r belongs through one of the formulas

$$MOD(r, P(m,n))$$
$$-P(m, INT[\frac{m-4+\sqrt{(m-4)^2+8(m-2)MOD(r,P(m,n))}}{2(m-2)}])$$

or

$$MOD(r, P(m,n))$$
$$-P(m, INT[\frac{-m+\sqrt{(m-4)^2+8(m-2)MOD(r,P(m,n))}}{2(m-2)}])$$

depending on whether relation (7) is satisfied or not. This completes the development of model for PS 7.

Example 2. Consider the case $r = 223$, $n = 5$, $m = 4$. We have

$$MOD(223, P(4,5))$$
$$-P(4, INT[\frac{4-4+\sqrt{(4-4)^2+8(4-2)MOD(223,P(4,5))}}{2(m-2)}])$$
$$= MOD(223,25) - P(4, INT[\frac{\sqrt{16 \cdot MOD(223,25)}}{4}])$$
$$= 23 - P(4,4) = 23 - 16 = 7.$$

Once can see (Figure 8.8) that room 223, indeed, belongs to the 7th storey.

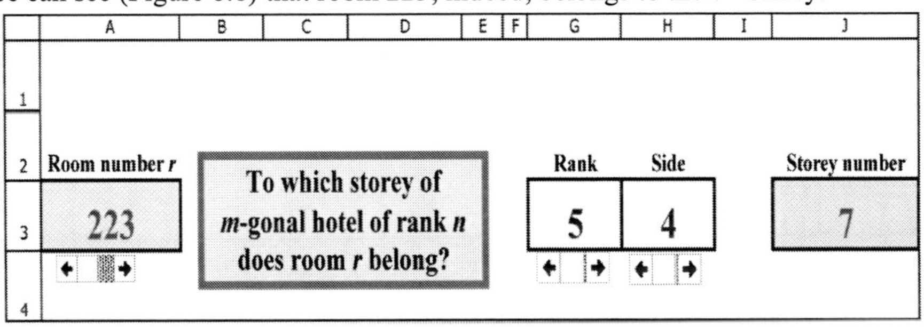

Figure 8.11. Spreadsheet modeling of PS 7.

In the case $r = 210$, $n = 7$ and $m = 5$ we have $MOD(210, P(5,7)) = MOD(210,70) = 0$ and therefore $(m-2)n-(m-3)\big|_{m=5,\,n=7} = 19$. Finally, in the case $r = 108$, $n = 6$, $m = 3$ we have

$$MOD(108, P(3,6)) - P(3, INT[\frac{-3+\sqrt{(3-4)^2+8(3-2)MOD(108,P(3,6))}}{2(3-2)}])$$
$$= MOD(108,21) - P(3, INT[\frac{-3+\sqrt{1+8 \cdot MOD(108,21)}}{2}])$$
$$= 3 - P(3, INT[\frac{-3+\sqrt{1+8 \cdot 3}}{2}]) = 3 - P(3,1) = 3 - 1 = 2.$$

The above three results can be confirmed by using the spreadsheet shown in Figure 8.11.

7. Introducing New Modeling Tools

PS 8. In which building of the pentagonal hotel of rank n does the j-th storey appear first?

Developing a model. Recall that a pentagonal hotel of rank n has $3n - 2$ stories. The meaning of the coefficient 3 is the increment in the number of stories when one moves from building to building. In turn, the difference $(3n - 2) - j$ represents the number of stories that separate the top storey (which for the first time appears in building n) and the one at level j.

If one moves from the top storey down by making a 3-storey descent at a time, each such descent can be counted as a move to the next building. Therefore, if storey j appears in building B first, then $n - B$ represents the number of buildings in the first block located to the right of building B. On the other hand, using the greatest integer function INT, this number of buildings can be described as $INT(\frac{3n-2-j}{3})$ whence

$$B = n - INT(\frac{3n-2-j}{3}) \qquad (8)$$

This completes the development of model for PS 8.

PS 9. In which building of the m-gonal hotel of rank n does the j-th storey appear first?

Developing a model. Let B represent the building number sought. Using equation (5) that describes the number of stories in an m-gonal hotel of rank n, formula (8) can be generalized to the form

$$B = n - INT(\frac{(m-2)n-(m-3)-j}{m-2}) \qquad (9)$$

thereby, giving an answer to PS 9. This completes the development of model for PS 8.

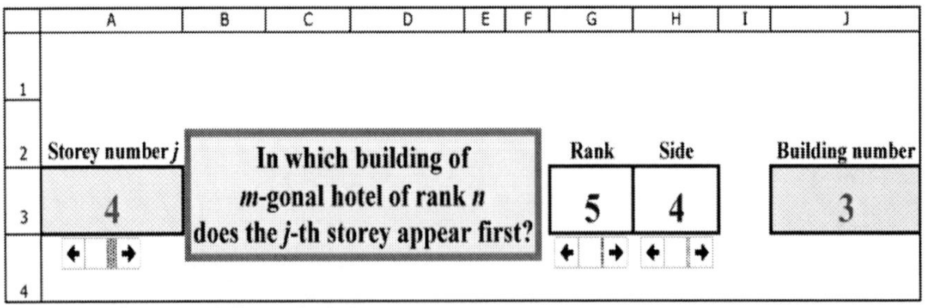

Figure 8.12. Spreadsheet modeling of PS 9.

Example 3. When $m = 4$, $n = 5$, and $j = 4$ formula (9) gives

$B = 5 - INT(\frac{2 \cdot 5 - 1 - 4}{2}) = 5 - 2 = 3$. This result can be confirmed by the blueprint of Figure 8.8 and the spreadsheet pictured in Figure 8.12.

8. Quadratic Functions as Tools of Summation

An interesting task in recognizing patterns and using quadratic functions as problem-solving tools is to define algebraically the sequence of room numbers that belong to the given storey as the function of building number *i*. To begin, consider the sequence of room numbers on the first storey of the first block of the pentagonal hotel of rank six, $f_1(i) = \{1, 2, 6, 13, 23, 36 \mid i = 1, ..., 6\}$ (Figure 8.3). This is a quadratic sequence as its second difference is a constant. Indeed, the first difference $\Delta_1 f_1 = \{1, 4, 7, 10, 13\}$ and $\Delta_2 f_1 = 3$. The same is true for all other sequences of room numbers (Figure 8.3):

$f_2(i) = \{3, 7, 14, 24, 37 \mid i = 2, ..., 6\}$, $\Delta_2 f_2 = 3$;

$f_3(i) = \{4, 8, 15, 25, 38 \mid i = 2, ..., 6\}$, $\Delta_2 f_3 = 3$.

Therefore, in order to find the coefficients *a*, *b*, *c* of the quadratic function $f_1(i) = ai^2 + bi + c$, where $f_1(1) = 1$, $f_1(2) = 2$, and $f_1(3) = 6$, one has to solve the system of equations

$$a + b + c = 1,\ 4a + 2b + c = 2,\ 9a + 3b + c = 6$$

Simple algebraic transformations yield $a = 1.5$, $b = -3.5$, and $c = 3$. (Alternatively, one can use the *fsolve* function of *Maple* [Figure 8.13] to solve this and similar systems of linear equations appearing below). That is, $f_1(i) = 1.5i^2 - 3.5i + 3$, where $i = 1, 2, ..., 6$. Then the function $f_{1,k}(i) = 1.5i^2 - 3.5i + 3 + 51k,\ k = 0, 1, 2, ...,\ i = 1, 2, ..., 6$, represents the sequence of room numbers on the first storey.

```
Storey1: fsolve({a+b+c=1, 4a+2b+c=2, 9a+3b+c=6}, {a, b, c})
                              {a=1.500000000, b=-3.500000000, c=3.}
Storey2: fsolve({a+b+c=3, 4a+2b+c=7, 9a+3b+c=14}, {a, b, c})
                              {a=1.500000000, b=-0.5000000000, c=2.}
Storey3: fsolve({a+b+c=4, 4a+2b+c=8, 9a+3b+c=15}, {a, b, c})
                              {a=1.500000000, b=-0.5000000000, c=3.}
Storey4: fsolve({a+b+c=5, 4a+2b+c=9, 9a+3b+c=16}, {a, b, c})
                              {a=1.500000000, b=-0.5000000000, c=4.}
Storey5: fsolve({a+b+c=10, 4a+2b+c=17, 9a+3b+c=27}, {a, b, c})
                              {a=1.500000000, b=2.500000000, c=6.}
Storey6: fsolve({a+b+c=11, 4a+2b+c=18, 9a+3b+c=28}, {a, b, c})
                              {a=1.500000000, b=2.500000000, c=7.}
Storey7: fsolve({a+b+c=12, 4a+2b+c=19, 9a+3b+c=29}, {a, b, c})
                              {a=1.500000000, b=2.500000000, c=8.}
Storey8: fsolve({a+b+c=20, 4a+2b+c=30, 9a+3b+c=43}, {a, b, c})
                              {a=1.500000000, b=5.500000000, c=13.}
Storey9: fsolve({a+b+c=21, 4a+2b+c=31, 9a+3b+c=44}, {a, b, c})
                              {a=1.500000000, b=5.500000000, c=14.}
Storey10: fsolve({a+b+c=22, 4a+2b+c=32, 9a+3b+c=45}, {a, b, c})
                              {a=1.500000000, b=5.500000000, c=15.}
```

Figure 8.13. Solving the system of three linear equations using *Maple*.

j	1	2, 3, 4	5, 6, 7	8, 9, 10
b	$-3.5+3\cdot 0=-3.5$	$-3.5+3\cdot 1=-0.5$	$-3.5+3\cdot 2=2.5$	$-3.5+3\cdot 3=5.5$

Figure 8.14. Relating coefficient *b* to the storey number *j*.

j	2	3	4	5	6	7	8	9	10
c	2	3	4	6	7	8	13	14	15
$j+p_k$	2 + 0	3 + 0	4 + 0	5 + 1	6 + 1	7 + 1	8 + 5	9 + 5	10 + 5

Figure 8.15. Relating coefficient *c* to the storey number *j*.

In much the same way, one can use *Maple* (Figure 8.13) to find that

$$f_2(i)=1.5i^2-0.5i+2\,,\,f_3(i)=1.5i^2-0.5i+3,\,f_4(i)=1.5i^2-0.5i+4\,.$$

As a result, the following quadratic functions can be constructed:

$$f_5(i)=1.5i^2+2.5i+6,\,f_6(i)=1.5i^2+2.5i+7\,,\;f_7(i)=1.5i^2+2.5i+8\,;$$
$$f_8(i)=1.5i^2+5.5i+13\,,\,f_9(i)=1.5i^2+5.5i+14\,,\,f_{10}(i)=1.5i^2+5.5i+15\,.$$

Analyzing the functions $f_j(i)$, j = 1, 2, 3, …, 10, one can recognize the following patterns regarding the dependence of coefficients a, *b*, and *c* on the storey number *j* as presented in the charts of Figure 8.14 (*b* versus *j*) and Figure 8.15 (*c* versus *j*).

First, it appears that regardless of *j* we have a = 1.5. Second, for each new triple of *j*-values, the coefficient *b* augments by three. However, the dependence of *c* on *j* is more complicated. Indeed, for the *j*-range [2, 4] the value of *j* increases by the pentagonal number of rank zero; for the *j*-range [5, 7] the value of *j* increases by the pentagonal number of rank one; for the *j*-range [8, 10] the value of *j* increases by the pentagonal number of rank two.

How can one relate the storey number *j* to the coefficients *b* and *c*? As the coefficient *b* does not change for a triple of numbers, one has to devise an operation, which transforms number *j* into one-third of the nearest multiple of three. With this in mind, consider the function *ROUND*(*x*, 0) which returns *x* rounded to the nearest integer and then define

$$b=-3.5+3\cdot ROUND(\frac{j}{3},0) \tag{10}$$

For example, when j = 10 we have

$$b=-3.5+3\cdot ROUND(\frac{10}{3},0)=-3.5+3\cdot 3=5.5\,.$$

Likewise, the value of coefficient *c* can be defined as follows

$$c = j + P[5, ROUND(\frac{j}{3},0) - 1]$$

or

$$c = j + \frac{[ROUND(\frac{j}{3},0) - 1][3 \cdot ROUND(\frac{j}{3},0) - 4]}{2} \quad (11)$$

For example, when $j = 16$ formulas (10) and (11) yield $b = -3.5 + 3 \cdot 5 = 11.5$ and $c = 16 + 4 \cdot 11/2 = 38$. These calculations are confirmed by the spreadsheet pictured in Figure 8.16 where cell A3 serves as input (storey number j) and formulas (10) and (11) are defined in cells C3 and D3, respectively. Therefore, $f_{16}(i) = 1.5i^2 + 11.5i + 38$.

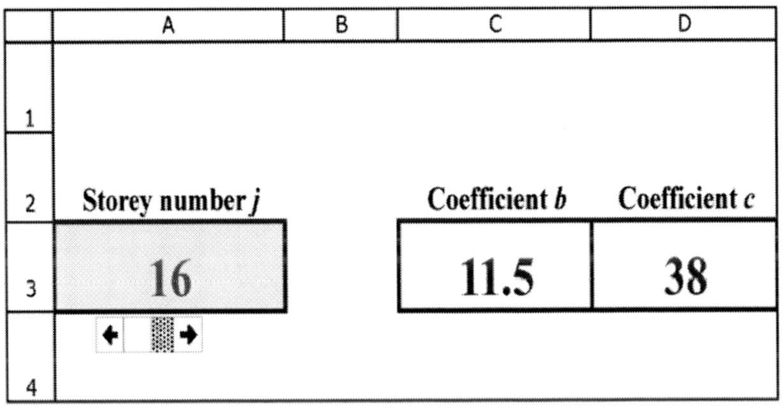

Figure 8.16. Calculating coefficients b and c using formulas (10) and (11).

9. Finding the Sum of Room Numbers that Belong to the Same Floor

Once again, the development of new models described by quadratic functions with coefficients defined by formulas (10) and (11), makes it possible to pose new problems (that is, to change context as a result of the development of new models).

PS 10. Find the sum of all room numbers on the 12th storey of the first block of the pentagonal hotel of rank 10.

Solution. Using formula (9) in the case $m = 5$, $n = 10$, and $j = 12$ yields $B = 10 - INT((28 - 12)/3) = 10 - 5 = 5$; that is, the 12th storey first appears in building 5. Therefore one has to find the sum of six room numbers (representing buildings 5 through 10). To this end, one has to construct the function $f_{12}(i) = 1.5i^2 + bi + c$, where $i = 1, 2, \ldots, 6$ (there are 6 buildings in a row having the 12th storey), and find the values of b and c. From formulas (10) and (11) it follows that

$$b = -3.5 + 3 \cdot ROUND(\frac{12}{3}, 0) = 8.5$$

and

$$c = 12 + \frac{[ROUND(\frac{12}{3}, 0) - 1][3 \cdot ROUND(\frac{12}{3}, 0) - 4]}{2} = 24.$$

Therefore, using formulas for the sums of consecutive counting numbers and their squares, the sum sought can be found as follows

$$\sum_{i=1}^{6} f_{12}(i) = \sum_{i=1}^{6}(1.5i^2 + 8.5i + 24) = 1.5\sum_{i=1}^{6} i^2 + 8.5\sum_{i=1}^{6} i + 24\sum_{i=1}^{6} 1$$
$$= 1.5\frac{6 \cdot (6+1) \cdot (2 \cdot 6 + 1)}{6} + 8.5\frac{6 \cdot (6+1)}{2} + 24 \cdot 6 = 459.$$

PS 11. Find the sum of all room numbers on the 12th storey of the first four blocks of the pentagonal hotel of rank 10.

Solution. As room numbers on a particular storey in each block are congruent to those in the first block, the sum sought can be found as follows

$$\sum_{i=1}^{6} f_{12}(i) + \sum_{i=1}^{6}[f_{12}(i) + 51] + \sum_{i=1}^{6}[f_{12}(i) + 2 \cdot 51] + \sum_{i=1}^{6}[f_{12}(i) + 3 \cdot 51]$$
$$= 4 \cdot \sum_{i=1}^{6} f_{12}(i) + 51 \cdot 6 \cdot (1 + 2 + 3) = 4 \cdot 459 + 36 \cdot 51 = 3672.$$

PS 12. Find the sum of all room numbers on the j-th storey of the first block of the pentagonal hotel of rank n.

Developing a model. One has to construct the quadratic function $f_j(i) = 1.5i^2 + bi + c$ where b and c are defined by formulas (10) and (11) and, according to formula (8),

$$i = 1, 2, \ldots, INT(\frac{3n - 2 - j}{3}) + 1.$$

In that way, the following sum serves as a model for PS 12:

$$\sum_{i=1}^{INT(\frac{3n-2-j}{3})+1} [1.5i^2 + (-3.5 + 3 \cdot ROUND(\frac{j}{3}, 0))i + j + \frac{(ROUND(\frac{j}{3}, 0) - 1)(3 \cdot ROUND(\frac{j}{3}, 0) - 4)}{2}] \quad (12)$$

In particular, when $j = 1$ expression (12) yields $\frac{n(n^2-2n+3)}{2}$—the sum of all room numbers on the first floor. Indeed, in that case

$$\sum_{i=1}^{INT(\frac{3n-2-1}{3})+1} [1.5i^2 + (-3.5 + 3\cdot ROUND(\frac{1}{3},0))i$$

$$+1+\frac{(ROUND(\frac{1}{3},0)-1)(3\cdot ROUND(\frac{1}{3},0)-4)}{2}]$$

$$=\sum_{i=1}^{n}[1.5i^2-3.5i+3]=\frac{3}{2}\sum_{i=1}^{n}i^2-\frac{7}{2}\sum_{i=1}^{n}i+3n=\frac{n(n+1)(2n+1)}{4}-\frac{7n(n+1)}{4}+3n$$

$$=\frac{n(n+1)}{4}(2n+1-7)+3n=n\frac{n^2-2n-3+6}{2}=\frac{n(n^2-2n+3)}{2}.$$

One can use *Maple* to calculate the value of expression (12) for different values of parameters n and j. An excerpt from *Maple* used in the case of PS 10 in shown in Figure 8.17.

Remark 2. A similar inquiry can be initiated by considering other values of m. This, however, would require new quadratic functions, the coefficients of which can be found similarly to the case $m = 5$.

10. Refining Old Models to Match New Context

Consider the blueprint pictured in Figure 8.18 where each block consists of sub-blocks of identical buildings forming the $1\times1, 2\times2, 3\times3$, and 4×4 squares. Just like in the case of polygonal hotels, questions about the number of rooms in a building, the relationship between room number and block number/floor number can be posed, and models for exploring these questions can be constructed.

> $j := 12$

$$j := 12$$

> $n := 10$

$$n := 10$$

> $ul1 := \frac{3\cdot n-2-j}{3}$

$$ul1 := \frac{16}{3}$$

> $ul := \text{round}(ul1)+1$

$$ul := 6$$

> $\sum_{i=1}^{ul}\left(1.5\cdot i^2+\left(-3.5+3\cdot\text{round}\left(\frac{j}{3}\right)\right)\cdot i+j+\left(\text{round}\left(\frac{j}{3}\right)-1\right)\cdot\left(3\cdot\text{round}\left(\frac{j}{3}\right)-4\right)\cdot 0.5\right)$

$$459.$$

Figure 8.17. The use of *Maple* in modeling PS 12.

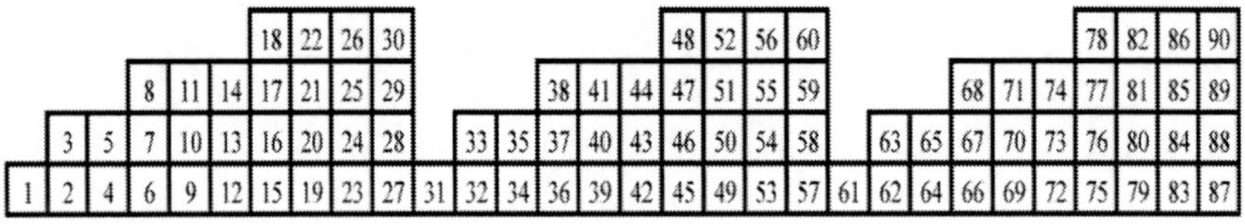

Figure 8.18. A blueprint of the first three blocks of the square type hotel of rank four.

Building range	1	2 – 3	4 – 6	7 – 10	11 – 15
Number of rooms	1	2	3	4	5

Figure 8.19. Relating building range to the number of rooms.

PS 13. How many rooms are in the 25th building of the 4-storied square-type hotel?

Solution. The new context does not require the design of a completely new model. Rather it requires the refinement of an already existing model. As shown in Figure 8.18, the number of buildings in a block, unlike the case of PS 1 (or PS 2), is not included in the givens (input). This number has to be found. To this end, the first step is to find the total number of buildings in a block. In the specific case of Figure 8.18, this number can be found as 1 + 2 + 3 + 4 = 10—the triangular number of rank four (see section 4). One can see that all buildings are arranged in blocks, ten buildings in each block. By analogy with PS 1 (or PS 2) one can map any given building number to the first block through the use of the MOD function. Thus the second step is to find MOD(25, 10) = 5. That is, the 5th and the 25th buildings have the same number of rooms. The blueprint shows that there are three rooms in the 5th building.

The question, however, remains how can one find the number of rooms given the building number that belongs to the 1st block? Unlike the case of PS 1 (or PS 2), the building number that belongs to the 1st block, in general, does not necessarily coincide with the number of rooms in this building. How can one formally (without using the blueprint) decide the number of rooms in the 5th building of the 1st block? One way to answer this question is to construct a chart that relates the range of building numbers to the number of rooms in that range. Such chart is shown in Figure 8.19.

One can recognize in the upper border of each building range a triangular number the rank of which appears in the cell immediately below and coincides with the number if rooms in a building from this range. The problem now has been reduced to the following one: *Given a positive integer, find the rank of the smallest triangular number that is greater or equal to this integer*. Recall that a more general problem was already explored in the context of PS 5, where the rank of the smallest m-gonal number was defined by formula (6). In particular, when $m = 3$ and $r_1 = 5$, formula (6) yields $CEILING(\frac{-1+\sqrt{41}}{2},1) = 3$—the rank of the smallest triangular number greater (or equal) to 5. This rank coincides with the number of rooms in the 5th building of the 4-storied square-type hotel. Alternatively, one can graph the equation $\frac{n(n+1)}{2} = 5$ for $n > 0$ (Figure 8.20) to have the graph in the form of a vertical line that crosses the x-axis in the interval [2, 3]. Once again, the upper integer bound of this segment, 3, represents the number of rooms in the 5th building, the result confirmed by the blueprint of Figure 8.18.

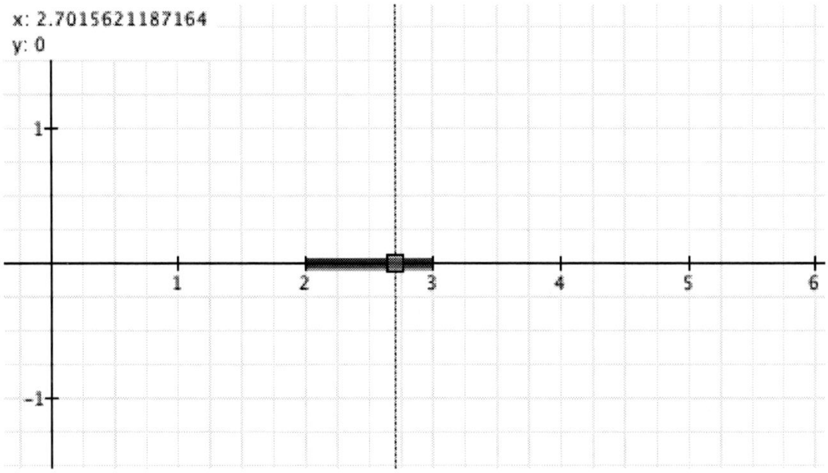

Figure 8.20. Finding the rank of the smallest triangular number greater than 5.

PS 14. How many rooms are in the *k*-th building of the n-storied square-type hotel?

Developing a model. Let *R*(n, *k*) represent the number of rooms sought. The first step in developing a model for finding *R*(*n*, *k*) is to find the total number of buildings in a block. This number is given by the sum $1+2+3+...+n=\frac{n(n+1)}{2}=t_n$—the triangular number of rank n(alternatively, the *n*-th partial sum of consecutive counting numbers). The second step is to map building *k* to the first block by evaluating $MOD(k,t_n)$ because the two buildings, *k* and $MOD(k,t_n)$, have the same number of rooms, assuming that $MOD(k,t_n)>0$. The third step is to find the positive root, n_1, of the equation $MOD(k,t_n)=t_n$, which is equivalent to the quadratic equation $n^2+n-2MOD(k,t_n)=0$ whence $n_1=\frac{-1+\sqrt{1+8MOD(k,t_n)}}{2}$.

Finally, *R*(*n*, *k*) can be found as the smallest integer greater or equal to n_1, that is,

$$R(n,k)=CEILING(\frac{-1+\sqrt{1+8MOD(k,t_n)}}{2},1),\ MOD(k,t_n)>0 \qquad (13)$$

When $MOD(k,t_n)=0$, room *k* is the last room in a block, thus *R*(*n*, *k*) = *n*. This model can be computerized by using the spreadsheet shown in Figure 8.21. In particular, there are five rooms in building 239 of the square type hotel of rank 9.

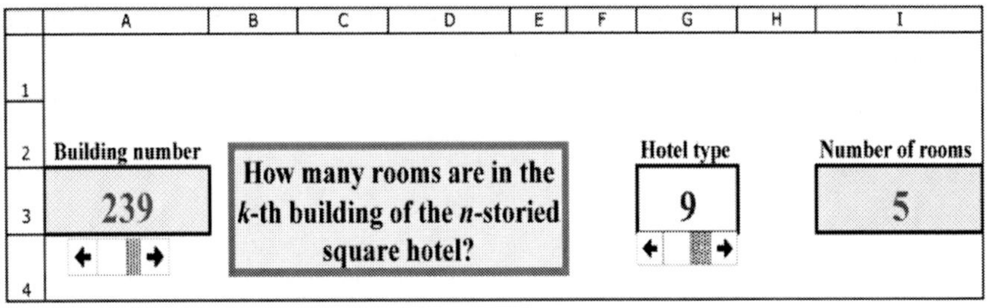

Figure 8.21. Computerization of PS 14.

PS 15. To which block of the 4-storied square-type hotel does room 65 belong?

Solution. The first step is to find the number of rooms in a block as the sum of four squares: $1^2+2^2+3^2+4^2=30$. The second step is to find the smallest number greater than 65 (room number) that is divisible by 30 (the number of rooms in a block) by evaluating *CEILING*(65, 30) = 90 which returns the smallest multiple of 30 greater than 65. The third step is to divide 90 by 30 to get 3—the block number sought.

Moving towards the development of a spreadsheet for evaluating the block number given room number and hotel type, one can formulate

PS 16. To which block of the *n*-storied square-type hotel does room r belong?

Developing a model. Let *B*(*r, n*) represent the block number sought. The first step is to find the number of rooms in an *n*-storied block as the sum of the first *n* squares of counting numbers: $1^2+2^2+3^2+...+n^2=\frac{n(n+1)(2n+2)}{6}$ (Chapter 1, formula [4]). The second step is to evaluate $CEILING(r,\frac{n(n+1)(n+2)}{6})$. The third step results in the completion of the model in the form of the formula

$$B(r,n)=CEILING(r,\frac{n(n+1)(2n+1)}{6})/[\frac{n(n+1)(2n+1)}{6}] \tag{14}$$

The spreadsheet shown in Figure 8.22 incorporates formula (14). In particular, room 678 of the 5-storied square-type hotel belongs to block 13.

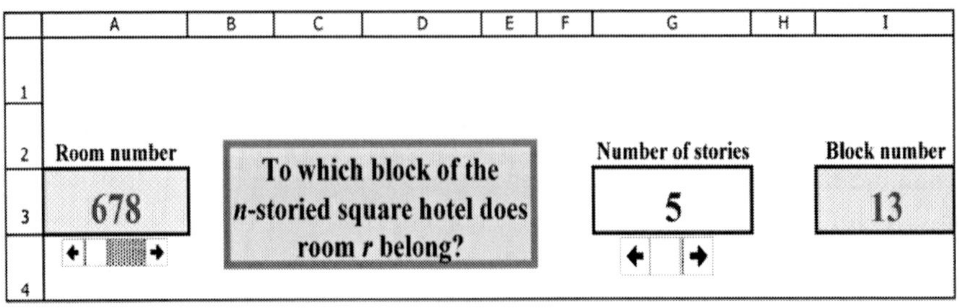

Figure 8.22. Computerization of PS 16.

11. INTERPRETING THE RESULTS OF COMPUTATIONAL EXPERIMENTS

Modeling activities described in this chapter can be extended towards the development of functional relationships for which there is no analytic formula. Such relationships are defined verbally. Consider PS 14. Given the building number, one can use a spreadsheet to explore the dependence of the number of rooms in this building on the number of stories in a square-type hotel. For example, using formula (13) one can find

$$R(25,4) = CEILING(\frac{-1+\sqrt{1+8MOD(25,t_4)}}{2},1)$$
$$= CEILING(\frac{-1+\sqrt{1+8MOD(25,10)}}{2},1) = CEILING(\frac{-1+\sqrt{41}}{2},1) = 3$$

and

$$R(25,5) = CEILING(\frac{-1+\sqrt{1+8MOD(25,t_5)}}{2},1)$$
$$= CEILING(\frac{-1+\sqrt{1+8MOD(25,15)}}{2},1) = CEILING(\frac{-1+\sqrt{81}}{2},1) = 4$$

That is, the 25-th buildings of the 4-storied and 5-storied square-type hotels have, respectively, three and four rooms. This shows that the number of rooms in a certain building of a square-type hotel is the function of the number of stories. How does this function (referred to below as $R(x,k)$, where variable x and parameter k denote, respectively, the number of stories and building number) behave for different values of k?

One can develop a spreadsheet environment that generates numerical (table) representations of this function for different building numbers. The results are shown in Figures 8.23-8.26 (a graphical representation of the case of the 180th building is shown in Figure 8.27). The following pattern in the behavior of the function for different building numbers can be observed: whatever the building number, one can always find a storey beginning from which the number of rooms stays constant. Furthermore, this invariant number of rooms coincides with the storey number where this number of rooms was observed first. How can this phenomenon be explained?

◇	A B C	D	E	F	G	H	I	J	K	L	M	N	O	P	Q	R	S	T	U	V	W	X	Y	Z	AA
1	150th building																								
2	Number of stories	2	3	4	5	6	7	8	9	10	11	12	13	14	15	16	17	18	19	20	21	22	23	24	25
3	Number of rooms	2	3	4	5	2	4	3	5	9	6	12	11	9	8	5	17	17	17	17	17	17	17	17	17

Figure 8.23. The behavior of $R(x, 150)$.

◇	A	B	C	D	E	F	G	H	I	J	K	L	M	N	O	P	Q	R	S	T	U	V	W	X	Y	Z	AA
1	160th building																										
2	Number of stories			2	3	4	5	6	7	8	9	10	11	12	13	14	15	16	17	18	19	20	21	22	23	24	25
3	Number of rooms			1	3	4	4	5	6	6	7	10	7	3	12	10	9	7	4	18	18	18	18	18	18	18	18

Figure 8.24. The behavior of $R(x, 160)$.

◇	A	B	C	D	E	F	G	H	I	J	K	L	M	N	O	P	Q	R	S	T	U	V	W	X	Y	Z	AA
1	170th building																										
2	Number of stories			2	3	4	5	6	7	8	9	10	11	12	13	14	15	16	17	18	19	20	21	22	23	24	25
3	Number of rooms			2	2	4	3	2	2	7	8	3	9	5	13	11	10	8	6	18	18	18	18	18	18	18	18

Figure 8.25.The behavior of $R(x, 170)$.

◇	A	B	C	D	E	F	G	H	I	J	K	L	M	N	O	P	Q	R	S	T	U	V	W	X	Y	Z	AA
1	180th building																										
2	Number of stories			2	3	4	5	6	7	8	9	10	11	12	13	14	15	16	17	18	19	20	21	22	23	24	25
3	Number of rooms			2	3	4	5	5	5	8	9	5	10	7	13	12	11	9	7	4	19	19	19	19	19	19	19

Figure 8.26.The behavior of $R(x, 180)$.

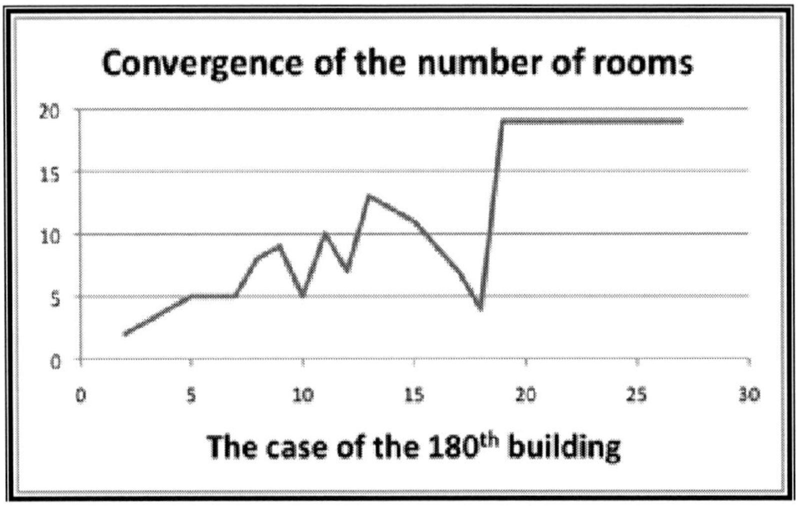

Figure 8.27. Graphical representation of the function $R(x, 180)$.

Recall that the number of rooms in the k-th building of the n-storied square-type hotel is equal to the value of the function

$$CEILING(\frac{-1+\sqrt{1+8MOD(k,t_n)}}{2},1)$$

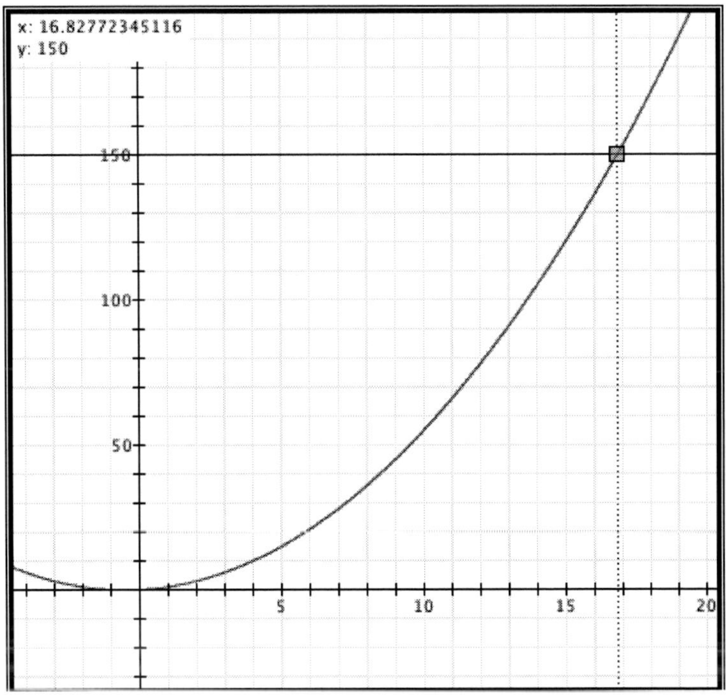

Figure 8.28. Solving the inequality $t_n > 150$.

For all n such that $t_n > k$, the equality $MOD(k,t_n) = k$ holds true. Indeed, the equality $5 = 0 \times 6 + 5$ indicates that dividing 6 into 5 results in the quotient zero and remainder 5. Likewise, dividing 7, 8, 9, … into 5 yields the same remainder, 5. What is the smallest value of n for which $t_n = \frac{n(n+1)}{2} > k$? In particular, what is the smallest value of n for which $t_n = \frac{n(n+1)}{2} > 150$? The last question can be answered by using computer graphics as shown in Figure 8.28 where the graph $y = \frac{n(n+1)}{2}$ intersects the line y = 150 at the point 16.83 so that n = 17 becomes the smallest positive integer for which $\frac{n(n+1)}{2} > 150$. Likewise, one can establish that the inequality $\frac{n(n+1)}{2} > 160$ holds true for $n \geq 18$, the inequality $\frac{n(n+1)}{2} > 170$ holds true for $n \geq 18$ also, and the inequality $\frac{n(n+1)}{2} > 180$ holds true for $n \geq 19$. This explains the modeling data of the spreadsheets pictured in Figures 8.23-8.26.

Note that one can observe a similar phenomenon for all models that are based on the use of the *MOD*(x,y) function due to the fact that *MOD*(x,y) = x when $y > x$. In particular, as the value of $\frac{n(n+1)}{2}$, where n is the number of stories in a counting-type hotel, becomes greater than either room number r or building number k, one has, respectively: room r stays on the same storey and the number of rooms in the k-th building does not change. Likewise, as pentagonal hotels become higher and higher, its first block may include more and more rooms thus, obviously making a room with sufficiently large number being a part of the first block. In other words, for any room number r one can find a pentagonal hotel for which this room resides in its first block. This shows how various phenomena observed through the interpretation of computational experiments with mathematical models have in fact rather obvious meanings.

12. Activity Set

Develop spreadsheet-based computational environments to answer the following questions.

1. How many rooms are in building 401 of the pentagonal hotel of rank 49?
2. How many rooms are in building 40 of the hexagonal hotel of rank 20?
3. To which building of the heptagonal hotel of rank 9 does room 113 belong?
4. To which storey of the octagonal hotel of rank 5 does room 205 belong?
5. In which building of the pentagonal hotel of rank 10 does the 18th storey appear first?
6. Find the sum of all room numbers on the 7th storey of the first block of the pentagonal hotel of rank 9.
7. How any rooms are in the 31st building of the 5-storied square-type hotel?
8. To which block of the 5-storied square-type hotel does room 222 belong?

Chapter 9

PROGRAMMING DETAILS

The best way to teach teachers is to make them ask and do what they, in turn, will make their students ask and do.

—Halmos (1975, p. 470)

1. INTRODUCTION

This chapter provides details of the programming of a number of spreadsheet environments used in this book. Learning to develop spreadsheet-based computational environments for the teaching of mathematics could be a part of a separate course on the educational technology for teachers. It has also been argued (Conference Board of the Mathematical Sciences, 2001) that teachers need experience in using spreadsheets across the whole spectrum of STEM-related undergraduate courses. With this in mind, it is assumed that the reader has basic familiarity with spreadsheets that includes the use of names (constants and variables) in spreadsheet formulas. This would allow one to master basic techniques of spreadsheet programming considered in this chapter. For a more advanced learning of the use of spreadsheets, a book by Neuwirth & Arganbright (2004) can recommended. Also, an open access online journal *Spreadsheets in Education* (http://epublications.bond.edu.au/ejsie/) can be used as an additional source of information about various uses of the software in the teaching of mathematics and other quantitatively based courses.

All computational environments discussed in this chapter are based on Excel 2008 (Mac)/2007(Windows) spreadsheets, but they work equally effective with Excel 2004 (Mac)/2003(Windows) versions. It should be noted that syntactic versatility of spreadsheets enables the construction of both visually and computationally identical environments using syntactically different formulas. Below, the notation (A1)→ will be used to present a formula defined in cell A1. Many spreadsheet formulas incorporate the conditional function =*IF*(_, _, _) which includes three parts: a condition, an action that must be taken if the condition is true, and an action that must be taken otherwise. Whenever appropriate, spreadsheet formulas include the names of constants and variables rather than cell references. The use of names facilitates one's comprehension of spreadsheet programming by giving meaning to the formulas involved. In many cases, a spreadsheet can be utilized as an agent of teachers' mathematical activities, thereby, supporting the recommendation, "prospective teachers need to understand that the use of technology for complicated computation does not eliminate the need for mathematical thinking but rather often raises a different set of mathematical problems" (Conference Board of Mathematical Sciences, 2001, p. 48).

2. Spreadsheets Used in Chapter 1

2.1. Programming Details for Figure 1.5

Cell A2 is slider-controlled and given the name n (size of the table).

(B2)→ =1; (C2)→ =IF(B2<n,1+B2," ")—replicated across row 2.

(A3)→ =1; (A4)→ =IF(A3<n,1+A3," ")—replicated down column A.

Highlight the range B2:M2 and define the name x; highlight the range A3:A14 and define the name y.

(B3)→ =IF(OR(B$2=" ",$A3=" ")," ",x*y)—replicated to cell M14. As a result, the spreadsheet generates the $n \times n$ multiplication table, $1 \le n \le 12$, the size of which is controlled by the slider attached to cell A2. In Figure 1.5, $n = 8$.

2.2. Programming Details for Figure 1.8

Cell A2 is slider-controlled and given the name n (size of the table).

(B2)→ =1; (C2)→ =IF(B2<n,1+B2," ")—replicated across row 2.

(A3)→ =1; (A4)→ =IF(A3<n,1+A3," ")—replicated down column A.

Highlight the range B2:M2 and define the name x; highlight the range A3:A14 and define the name y.

(B3)→ =IF(OR(B$2=" ",$A3=" ")," ",x*y)—replicated to cell M14. Highlight the range B3:M14. Open the dialogue box of Conditional Formatting (Format Menu).

Condition 1. Formula Is

AND(SQRT(B$2*$A3)=INT(SQRT(B$2*$A3)),MOD(B$2*$A3,2)=0, B$2=$A3). Chose format. As a result, the spreadsheet generates the $n \times n$ multiplication table, $1 \le n \le 12$, with even products that consist of two equal factors highlighted. In Figure 1.8, $n = 10$. To display the highlighted products only, the spreadsheet can be modified as follows:

(B3)→ =IF(OR(B$2=" ",$A3=" ")," ",IF(AND(SQRT(B$2*$A3) =INT(SQRT(B$2*$A3)),MOD(B$2*$A3,2)=0,B$2=$A3),x*y," "))—replicated to cell M14. The formula =SUM(B3:M14) generates the sum of the displayed products.

2.3. Programming Details for Figure 1.9

Cell A2 is slider-controlled and given the name n (size of the table).

(B2)→ =1; (C2)→ =IF(B2<n,1+B2," ")—replicated across row 2.

(A3)→ =1; (A4)→ =IF(A3<n,1+A3," ")—replicated down column A.

Highlight the range B2:M2 and define the name x; highlight the range A3:A14 and define the name y.

(B3)→ =IF(OR(B$2=" ",$A3=" ")," ",x*y)—replicated to cell M14. Highlight the range B3:M14. Open the dialogue box of Conditional Formatting (Format Menu). Condition 1. Formula Is

AND(SQRT(B$2*$A3)=INT(SQRT(B$2*$A3)),MOD(B$2*$A3,2)>0, B$2=$A3). Chose format. As a result, the spreadsheet generates the $n \times n$ multiplication table, $1 \le n \le 12$, with odd products that consist of two equal factors highlighted. In Figure 1.9, n = 9. To display the highlighted products only, the spreadsheet can be modified as follows:

(B3)→ =IF(OR(B$2=" ",$A3=" ")," ",IF(AND(SQRT(B$2*$A3) =INT(SQRT(B$2*$A3)),MOD(B$2*$A3,2)>0,B$2=$A3),x*y," "))—replicated to cell M14. The formula =SUM(B3:M14)generates the sum of the displayed products.

2.4. Programming Details for Figures 1.11 and 1.12

Cell A2 is slider-controlled and given the name n (size of the table).

(B2)→ =1; (C2)→ =IF(B2<n,1+B2," ")—replicated across row 2.

(A3)→ =1; (A4)→ =IF(A3<n,1+A3," ")—replicated down column A.

Highlight the range B2:M2 and define the name x; highlight the range A3:A14 and define the name y.

(B3)→ =IF(OR(B$2=" ",$A3=" ")," ",x*y)—replicated to cell M14. Highlight the range B3:M14. Open the dialogue box of Conditional Formatting (Format Menu). Condition 1. Formula Is

AND(MOD(B$2,2)=0,MOD($A3,2)=0). Chose format. As a result, the spreadsheet generates the $n \times n$ multiplication table, $1 \le n \le 12$, with products that consist of both even factors highlighted. In Figure 1.11, n = 9; in Figure 1.12, n = 10.

To display the highlighted products only, the spreadsheet can be modified as follows: (B3)→

=IF(OR(B$2=" ",$A3=" ")," ",IF(AND(MOD(B$2,2)=0,MOD($A3,2)=0), x*y," "))—replicated to cell M14. Alternatively, the spreadsheet displays products in non-highlighted cells when (B3)→

=IF(OR(B$2=" ",$A3=" ")," ",IF(AND(MOD(B$2,2)=0,MOD($A3,2) =0)," ",x*y))—replicated to cell M14.

In any case, the formula =SUM(B3:M14) generates the sum of the displayed products.

2.5. Programming Details for Figures 1.10 and 1.13

Cell A2 is slider-controlled and given the name n (size of the table).

(B2)→ =1; (C2)→ =IF(B2<n,1+B2," ")—replicated across row 2.

(A3)→ =1; (A4)→ =IF(A3<n,1+A3," ")—replicated down column A.

Highlight the range B2:M2 and define the name x; highlight the range A3:A14 and define the name y.

(B3)→ =IF(OR(B$2=" ",$A3=" ")," ",x*y)—replicated to cell M14. Highlight the range B3:M14. Open the dialogue box of Conditional Formatting (Format Menu). Condition 1. Formula Is

AND(MOD(B$2,2)>0,MOD($A3,2)>0). Chose format. As a result, the spreadsheet generates the $n \times n$ multiplication table, $1 \leq n \leq 12$, with products that consist of both odd factors highlighted. In Figure 1.10, $n = 10$; in Figure 1.13, $n = 9$. To display the highlighted products only, the spreadsheet can be modified as follows: (B3)→

=IF(OR(B$2=" ",$A3=" ")," ",IF(AND(MOD(B$2,2)>0,MOD($A3,2)>0), x*y," "))—replicated to cell M14. Alternatively, the spreadsheet displays products in non-highlighted cells when (B3)→

=IF(OR(B$2=" ",$A3=" "),"",IF(AND(MOD(B$2,2)>0,MOD($A3,2) >0)," ",x*y))—replicated to cell M14.

In any case, the formula =SUM(B3:M14) generates the sum of the displayed products.

3. Spreadsheets Used in Chapter 3

3.1. Programming Details for Figures 3.1 and 3.2

Cells B1, B3, B5 are slider-controlled and given the names n, a, b, respectively.

(N1)→ =0; (O1)→ =IF(N1=" "," ",IF(N1<INT(n/a),N1+1," "))—replicated across row 1.

(M2)→ =0; (M3)→ =IF(M2=" "," ",IF(M2<INT(n/b),M2+1," "))—replicated down column M.

(N2)→ =IF(OR(N$1=" ",$M2=" ")," ",IF(n=a*N$1+b*$M2,n," "))—replicated across rows and down columns. In Figure 3.1, $n = 18$; in Figure 3.2, $n = 24$.

3.2. Programming Details for Figures 3.3 and 3.4

Cells B1, B3, B5, E7, E8 are slider-controlled and given the names n, a, b, xmin, ymin, respectively.

(N1)→ =xmin; (O1)→ =IF(N1=" "," ",IF(N1<INT((n-ymin*b)/a),N1+1, " "))—replicated across row 1.

(M2)→ =ymin; (M3)→ =IF(M2=" "," ",IF(M2<INT((n-xmin*a)/b),M2+1," ")—replicated down column M.

(N2)→ =IF(OR(N$1=" ",$M2=" ")," ",IF(n=a*N$1+b*$M2,n," "))—replicated across rows and down columns. In Figure 3.3, $n = 18$, xmin = 0, ymin = 2; in Figure 3.2, $n = 24$, xmin = 3, ymin = 0.

3.3. Programming Details for Figure 3.5

Cell B1 is slider-controlled and given the name n.

(D2)→ =n+1; (E2)→ =IF(D2<2*n,D2+1," ")—replicated across row 2; highlight the range covered by the replicated formula and define the name x;

(C3)→ =2*n; (C4)→ =IF(C3<n*(n+1),C3+1," ")—replicated down column C; highlight the range covered by the replicated formula and define the name y;

(D3)→ =IF(OR(x=" ",y=" "), " ",IF(1/n=1/x+1/y,n," ")—replicated across rows and down columns. In Figure 3.5, $n = 6$.

3.4. Programming Details for Figure 3.6

Cell B3 is slider-controlled and given the name n.

(I2)→ =n+1; (L2)→ =n*(n+1);

(C4)→ =n+2; (D4)→ =IF(C4<2*n,C4+1," ")—replicated across row 2; highlight the range covered by the replicated formula and define the name x;

(B5)→ =2*n; (B6)→ =IF(B5<n*(n+2)/2,B5+1," ")—replicated down column B; highlight the range covered by the replicated formula and define the name y;

(D3)→ =IF(OR(x=" ",y=" "), " ",IF(1/n=1/x+1/y,n," ")—replicated across rows and down columns. In Figure 3.6, $n = 6$.

3.5. Programming Details for Figure 3.9.

Cell B2 is slider-controlled and given the name n.

(B3)→ =3*n; (B4)→ =2*n*(n+1); (B5)→=(n*(n+1)+1)*n*(n+1);

(E1)→ =n+1; (F1)→ =IF(E1<3*n,E1+1," ")—replicated across row 1; highlight the range covered by the replicated formula and define the name x;

(D2)→ =2*n+1; (D3)→ =IF(D2<2*n*(n+1),D2+1," ")—replicated down column D; highlight the range covered by the replicated formula and define the name y;

(E2)→ =IF(OR(x=" ",y=" ")," ",IF(x*y=n*(x+y)," ",IF(AND(x<=y,
INT(n*x*y/(x*y-n*(x+y)))=n*x*y/(x*y-n*(x+y)),
n*x*y/(x*y-n*(x+y))>=y),n*x*y/(x*y-n*(x+y))," ")))—replicated across rows and down columns. In Figure 3.9, $n = 3$.

4. Spreadsheets Used in Chapter 8

4.1. Programming Details for Figure 8.6

Cells A3 and G3 are slider-controlled and given the names B and n, respectively.

(I3)→ =IF(MOD(B,n)=0, 3*n-2, 3*MOD(B,n)-2).

4.2. Programming Details for Figure 8.9

Cells A3, G3, and H3 are slider-controlled and are given the names B, n, and m, respectively.

(J3)→ =IF(MOD(B,n)=0,(m-2)*n-(m-3),(m-2)*MOD(B,n)-(m-3)).

4.3. Programming Details for Figure 8.10

Cells A3, G3, and H3 are slider-controlled and given the names r_, n, and m, respectively.

(J3)→ =CEILING((m-4+SQRT((m-4)^2+8*(m-2)*MOD(r_,
(0.5*n*(n-1)*(m-2)+n))))/(2*m-4),1)+n*INT(r_/(0.5*n*(n-1)*(m-2)+n)).

4.4. Programming Details for Figure 8.11

Cells A3, G3, and H3 are slider-controlled and given the names r_, n, and m, respectively.

(A1)→ =MOD(r_,0.5*n*(n-1)*(m-2)+n)—hidden from view;

(B1)→ =IF(INT((m-4+SQRT((m-4)^2+8*(m-2)*A1))/(2*m-4))
=(m-4+SQRT((m-4)^2+8*(m-2)*A1))/(2*m-4),(-m+SQRT((m-4)^2
+8*(m-2)*A1))/(2*m-4),INT((m-4+SQRT((m-4)^2+8*(m-2)*A1))/(2*m-4)))—hidden from view;

(J3)→ =IF(A1>0,A1-0.5*B1*(B1-1)*(m-2)-B1,(m-2)*n-(m-3)).

4.5. Spreadsheet Programming for Figure 8.12

Cells A3, G3, and H3 are slider-controlled and given the names j, n, and m, respectively.

(J3)→ =n-INT(((m-2)*n-(m-3)-j)/(m-2)).

4.6. Spreadsheet Programming for Figure 8.16

Cell A3 is slider-controlled and given the name j.

(C3)→ =-3.5+3*ROUND(j/3,0);

(D3)→ =j+0.5*(ROUND(j/3,0)-1)*(3*ROUND(j/3,0)-4).

4.7. Spreadsheet Programming for Figure 8.21

Cells A3 and G3 are slider-controlled and given the names k and n, respectively.

(I3)→ =IF(MOD(k,n*(n+1)/2)=0, n,
CEILING(0.5*(-1+SQRT(1+8*MOD(k,n*(n+1)/2))),1)).

4.8. Spreadsheet Programming for Figure 8.22

Cells A3 and G3 are slider-controlled and given the names r_ and n, respectively.

(I3)→ =CEILING(r_,n*(n+1)*(2*n+1)/6)/(n*(n+1)*(2*n+1)/6).

REFERENCES

Abramovich, S. (1995). Technology for deciding the convergence of series. *International Journal of Mathematical Education in Science and Technology*, *26*(3), 347-366.

Abramovich, S. (2003). Spreadsheet-enhanced problem solving in context as modeling. *Spreadsheets in Education (eJSiE)*, *1*(1), pp.1-17. Available at: http://epublications.bond.edu.au/ejsie/vol1/iss1/1.

Abramovich, S. (2006). Spreadsheet modelling as a didactical framework for inequality-based reduction. *International Journal of Mathematical Education in Science and Technology*, *37*(5), 527-541.

Abramovich, S. (2010). *Topics in Mathematics for Elementary Teachers: A Technology-Enhanced Experiential Approach*. Charlotte, NC: Information Age Publishing.

Abramovich, S., and Sugden, S. (2004). Spreadsheet conditional formatting: an untapped resource for mathematics education. *Spreadsheets in Education (eJSiE)*, *1*(2), pp. 85-105. Available at: http://epublications.bond. edu.au/ejsie/vol1/iss2/3.

Alexandrov, A. D. (1963). A general view of mathematics. In A. D. Alexandrov, A. N. Kolmogorov, and M. A. Lavrent'ev (Eds), *Mathematics: Its Content, Methods, and Meaning* (pp. 1-64). Cambridge, MA: The MIT Press.

Bradley, R. (2005). *Ritual and Domestic Life in Prehistoric Europe*. New York, Routledge.

Conference Board of the Mathematical Sciences.(2001).*The Mathematical Education of Teachers*. Washington, DC: The Mathematical Association of America.

Cooligde, J. L. (1963). *A History of Geometrical Methods*. New York: Dover.

Davis, P. J. (1985). What do I know? A study of mathematical self-awareness. *College Mathematics Journal*, *16*(1), 22-41.

Davis, P. J. (1993). *Spirals: From Theodorus to Chaos* (with contributions by W. Gautshi and A. Iserles). Wellsley, MA: AK Peters.

Descartes, R. (1965). *A Discourse on Method*. London: J. M. Dent and Sons.

Dunham, W. (1999). *Euler. The Master of Us All*. Washington, DC: The Mathematical Association of America.

Gershenfeld, N. (2005). *Fab: The Coming Revolution on Your Desktop—from Personal Computers to Personal Fabrication*. New York: Basic Books.

Hadamard, J. (1996). *The Mathematician's Mind: The Psychology of Invention in the Mathematical Field*. Princeton, NJ: Princeton University Press.

Halmos, P. R. (1975). The teaching of problem solving. *American Mathematical Monthly,* 82(5), 466-470.

Harrison, J. (2008). Formal proof—theory and practice. *Notices of the American Mathematical Society*, *55*(11), 1395-1406.

Hersh, R. (1993). Proving is convincing and explaining. *Educational Studies in Mathematics*, *24*(4), 389-399.

Hoffman, P. (1998). *The Man Who Loved Only Numbers: The Story of Paul Erdös and the Search for Mathematical Truth*. New York: Hyperion.

Jacobsen, L., Thron, W. J., and Waadeland, H. (1989). Julius Worpitzky, his contributions to the analytic theory of continued fractions and his times. In L. Jacobsen (Ed.), *Analytic Theory of Continued Fractions III* (pp. 25-47). Berlin: Springer-Verlag.

Katehi, L., Pearson, G., and Feder, M. (2009). *Engineering in K-12 Education*. Washington, DC: The National Academies Press.

Klein, F. (1979). *Development of Mathematics in the 19th Century*. Brookline, MA: Math Sci Press.

Kline, M. (1980). *Mathematics: The Loss of Certainty*. New York: Oxford University Press.

Kline, M. (1985). *Mathematics for the Non-Mathematician*. New York: Dover.

Korovkin, P. P. (1961). *Inequalities*. New York: Blaisdell Publishing Company.

Krantz, S. G. (2009). *A Guide to Real Variables*. Washington, DC: The Mathematical Association of America.

Langtangen, H. P., and Tveito, A. (2001). How should we prepare the students of science and technology for a life in the computer age? In B. Engquist and W. Schmid (Eds), *Mathematics Unlimited—2001 and Beyond* (pp. 809-825). New York: Springer.

National Council of Teachers of Mathematics. (1989). *Curriculum and Evaluation Standards for School Mathematics*. Reston, VA: Author.

National Council of Teachers of Mathematics.(2000). *Principles and Standards for School Mathematics*. Reston, VA: Author.

Neuwirth, E., and Arganbright, D. (2004). *Mathematical Modeling with Microsoft® Excel*. Belmont, CA: Thomson Learning.

Oxford Dictionary of Scientific Quotations. (2005). Edited by W. F. Bynum and R. Porter. New York: Oxford University Press.

Piaget, J. (1954). *The Construction of Reality in the Child*. New York: Basic Books.

Pólya, G. (1981). *Mathematical Discovery: On Understanding, Learning, and Teaching Problem Solving* (combined edition). New York: Wiley.

Pólya, G. (1973). *How to Solve It*? Princeton, NJ: Princeton University Press.

Pólya, G. (1954). *Induction and Analogy in Mathematics* (volume 1). Princeton, NJ: Princeton University Press.

Reid, C. (1963). *A Long Way from Euclid*. New York: Thomas Y. Crowell Company.

Solomon, H. (1978). *Geometric Probability*. Philadelphia, PA: Society for Industrial and Applied Mathematics.

Thibaut, G. (1984). *Mathematics in the Making in Ancient India*. Calcutta: K P Bagghi and Company.

Van der Waerden, B. L. (1961). *Science Awakening*. New York: Oxford University Press.

Weisstein, E. W. (1999). *CRC Concise Encyclopedia of Mathematics*. Boca Raton, FL: Chapman and Hall/CRC.

INDEX